AMAZONIA:
VOICES FROM THE RAINFOREST

A Resource & Action Guide

Editor
Angela Gennino

Publishers
Rainforest Action Network
Amazonia Film Project

Co-sponsors
South and Meso-American Indian Information Center
International Rivers Network

Writers
Monti Aguirre, Angela Gennino, Brent Millikan,
Maria Amália Souza, Glenn Switkes

Contributors
Beto Borges, Alberto Chirif, François Correa, Wade Davis, Scott Desposato,
John Frichione, Roberto Lizarralde, Juergen Riester,
Mariano Useche, Bill Walker, Lucilene and Ted Whitesell

Graphic Design
Kara Adanalian

Typesetting and Production
Steve Barton

Cartography
Anneke Vonk

Flower Illustrations
Kathy Flynn

Cover Illustration
Belinda van Valkenburg

Delta College Library

Special Thanks
Damien Foundation, Anne Withy, Seven Springs Foundation, Martha Lyddon,
Barbara Keller, Cultural Survival, Andrew Martin, Onaway Trust, Judy Donald.

©1990 Rainforest Action Network
301 Broadway, Suite A, San Francisco, 94133 U.S.A. (415) 398-4404

Table of Contents

Introduction	1
Amazonian Almanac	2
Native Amazonians	4
Forest Gatherers	10
Colonists	16
Support Groups	22
Directory	
South America	
COICA	27
Bolivia	28
Brazil	32
Colombia	51
Ecuador	54
Guyana	57
Peru	58
Suriname	62
Venezuela	62
North America	
Canada	65
United States	66
Europe	
Austria	77
Belgium	77
Denmark	78
England	78
Hungary	81
Italy	81
Netherlands	82
Norway	82
Spain	83
Sweden	83
Switzerland	83
West Germany	84
Asia/Pacific	
Australia	85
India	85
Japan	85
Malaysia	85
Recommended Books and Films	87
Index	89

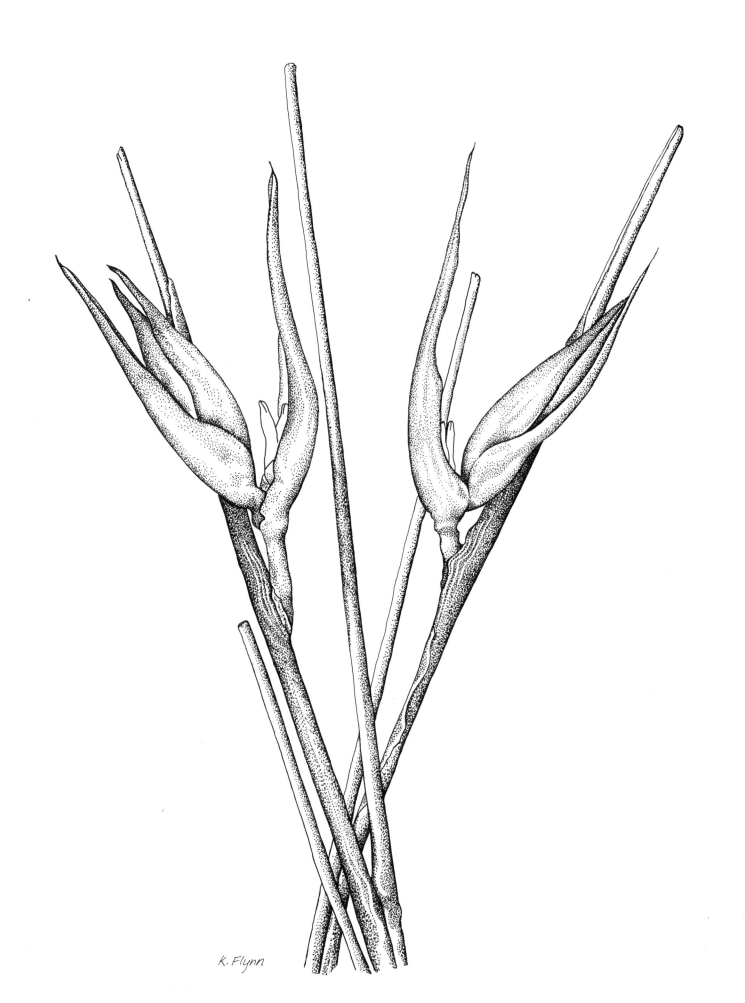
K. Flynn

Introduction

For the last few years, many of us have watched in disbelief as horrific scenes of rainforest annihilation filled the news media. Photographs of fires devouring canopy trees rage on the color pages of *Time*, *National Geographic*, and *Rolling Stone*. On TV, scientists paddle into the reservoir of a new hydroelectric dam to rescue terrified wildlife from the drowning rainforest. Newspapers report that Chico Mendes was gunned down in his own house for trying to stop the destruction. Even from outer space, satellite sequences show a steady expansion of scarred landscape: deforestation out of control.

All of these images are from the Amazon, the river basin containing the Earth's largest contiguous rainforest and 20 percent of the planet's fresh water. Biologists and ecologists say there are more than 23,000 species of life in 2.5 square kilometers of the Amazon—the greatest biological diversity on Earth. And yet some 10,360 square kilometers of the Amazon are being torched each month, destroying an area the size of Pennsylvania each year. Worse yet, destroying the forest destroys the people who live there. During this century, in Brazil alone, one Indian culture has disappeared each year. With each lost culture, ethnobotanists say, centuries of accumulated knowledge of the forest are also lost forever.

What scientists can't so easily tell us is how to save these areas and support the people of the rainforest. In the torching of the Amazon, there is no Darth Vader, no one institution, corporation, or government figure to write or to boycott. And the harder we look for the culprits, the closer we get to ourselves: We are the consumers of tropical products; we are the taxpayers whose governments and multilateral development banks finance ecological destruction and social and economic chaos. It is an activists' nightmare, with the same questions asked repeatedly: "Where do we start fighting? What is the most effective strategy? Who is interested in a rainforest thousands of kilometers away? Are there any groups in the Amazon doing anything? Who are they?"

Hundreds of South American organizations are defending the Amazon and its inhabitants, but most do not even know there are people in the North deeply concerned about the issue. Those who do know have their own questions: "How can we get support around the world? Is it true that North American and European organizations get millions of dollars to save the rainforests? How do they spend all that money? Who are the people behind these groups like the Rainforest Action Network? What are other organizations doing in the rest of the Amazon?"

This Action Guide was born of the need to address just such questions. It is designed as an organizing tool to help people and groups who are about to plunge into Amazonian work for the first time. The bulk of the guide is a directory of more than 250 organizations who are working on the issue, from the most remote parts of the jungle to the lobbying offices of Washington and London.

But the Action Guide is more than addresses and fax numbers. Unless the social and political complexities of life in the Amazon are understood, all the international campaigns and good intentions in the world will not keep a single tree standing. This guide, then, spotlights the grassroots movements of Amazon's indigenous peoples, rubber tappers, and other forest dwellers. No other people have worked as hard and as long to defend the forest. No other people understand as well the meaning—and importance—of sustainable development and conservation.

With this in mind, the Action Guide staff, for the first time, asked Amazonian grassroots organizations what they wanted the world to know about their goals, strategies, and special needs to make their work more effective. The information for more than 100 profiles was compiled from a series of questionnaires and from interviews conducted by Monti Aguirre and Glenn Switkes during their filming of *AMAZONIA: Voices from the Rainforest*, a documentary for which this publication is intended as a companion guide. Some 150 activist organizations working on Amazonian issues in Latin America, Europe, North America, and in the Asia/Pacific region, as well as resource groups of funders, researchers, and technical specialists responded to similar questionnaires.

Although the questionnaire's response rate was impressive, the guide is by no means complete. As we go to press, new organizations are forming every day and questionnaires, often mailed from remote villages, are still trickling into the offices of the Rainforest Action Network. Through the Portuguese and Spanish editions of the guide, still to be published, we will make an ongoing effort to update the directory and encourage greater collaboration between the peoples of the rainforest and activists around the world.

Amazonian Almanac

▼ Rank of the Amazon among the largest and longest rivers in the world: 1

▼ Rank of the Amazon basin among the Earth's largest contiguous rainforests: 1

▼ Proportion of the area of the Amazon to all rainforests on Earth: 1:3

▼ Ratio of the size of the Amazon basin to the continental United States: 1:1

▼ Vertical drop, in meters, from the Amazon's Andean headwaters in the first 960 kilometers: 4,500

▼ In the next 2,560 kilometers to the mouth: 78

▼ Total length, in kilometers, of the Amazon River: 6720

▼ Kilometers of navigable river in the Amazon basin: 80,000

▼ Percentage of Earth's fresh water flowing through the Amazon: 20

▼ Ratio of the Amazon's flow to that of the Mississippi: 11:1

▼ Cubic meters of water flowing through the Amazon's mouth per second: 198,000

▼ Days required for the Amazon's flow to exceed the yearly flow from the Thames: 1

▼ Hours required for the Amazon's flow to fill a hypothetically drained Lake Ontario: 3

▼ Percentage of the Amazon basin made up of floodplain: 3

▼ Of interior forest: 97

▼ Average rainfall, in centimeters, in the Amazon each year: 254

▼ Percentage of Amazonian trees whose roots extend less than 10 centimeters below the surface: 90

▼ Percentage of all known plant and animal species found in the Amazon: 30

▼ Number of species that may be found in a 2.5 square hectares of Amazonian rainforest: 23,000

▼ Number of species of trees that can be found in one hectare of Amazonian rainforest: 300

▼ In a California forest: 5 to 10

▼ Number of plant species in all of New England: 1,200

▼ In the Amazon: 80,000

▼ Estimated number of insects living in the Amazon: 30 million

▼ Number of Bolivian tree stumps found to house more ant species than in all of Great Britian: 1

▼ Rank of the Rio Negro among world rivers in diversity of fish species: 1

▼ Rank of Colombia among nations of the world in diversity of bird species: 1

▼ Species of butterflies found in Peru: 1,450

▼ In all of North America: 730

▼ Number of hectares of the Amazon lost, on average, to deforestation per minute: 20

▼ Per day: 30,000

▼ Per year: 8 million

▼ Total area deforested to date, in square kilometers: 356,200

▼ Percentage of total lost since 1960: 75

▼ Estimated percentage of entire rainforest lost to date: 7 to 20

▼ Size, in square kilometers, of one fire set in 1974 by Volkswagen to clear Brazilian forest: 10,000

▼ Rank of VW fire among largest man-made fires in world: 1

▼ Number of fires visible by satellite on a given day in Rondônia and Mato Grosso, Brazil, each July to September: 8,000

▼ Metric tons of carbon dioxide emitted by each fire: 4,500

▼ Of carbon monoxide: 750

▼ Of methane: 25

▼ Of total particulates emitted each burning season: 10 million

▼ Percentage of carbon dioxide emissions from worldwide deforestation that comes from the Amazon: 33

Native Amazonians

The Amazon, named for the legendary warrior women who turned back a terrified expedition of conquistadores some 500 years ago, remains a place of conflict between its original inhabitants and outsiders who come in conquest. Some indigenous people still don war paint and sharpen their arrows to raid the camps of settlers, ranchers, and miners. But many have learned that political pressure is a key weapon in the fight for the integrity of their lands and culture.

At the time of the arrival of the first Europeans, Amazonia was home to hundreds of distinct and complex native cultures. By some estimates, 8 million or more people were living in the Amazon when the whites "discovered" it. The invasion and brutal wars of conquest brought the slaughter of entire civilizations.

Guns were not the colonists' only weapons. Contagious diseases such as smallpox and measles, for which Indian people had no natural antibodies, killed millions. Even today, disease takes a heavy toll. Within just three years of their first contact in 1969, half of the Suruí people of Rondônia had died of diseases traditional tribal healers could not treat. Itabira, the Suruí chief, was in despair. Two years ago, after the death of his child from tuberculosis, he sent an open letter to the outside world. "I feel like crying out to tell of our sufferings," wrote Itabira, "but I have no means to leave here; no means to let other people in Brasília know. ...We have no money to go to other places. ... We would like to have your help in our fight."

Suruí Indian digging for grubs in the rainforest.
Photo: ©Aguirre/Switkes

Calha Norte: The Military Keeps a Tight Grip on the Amazon

Just two months after the Brazilian military transferred power to civilian rule in 1985, it quietly mapped out Calha Norte (the Northern Basin Project). This ambitious plan was designed to bring the entire 6,400-kilometer-long northern region firmly under military control by opening up Indian-occupied areas to dozens of military installations and mining projects.

For the military, security and development of the Amazon have always gone hand in hand. It was the military dictatorship that built the Trans-Amazon Highway in the 1970s; offered generous tax breaks to ranchers and other investors in the Amazon; started Polonoroeste, which brought more than 1 million migrants to the state of Rondônia; and initiated Calha Norte's forerunner, the Northern Perimeter Road project. The road (BR 210) wiped out vast sections of the rainforest mainly inhabited by 20,000 Yanomami, the largest traditional Indian group in the Americas.

After the return to civilian rule, the military retained control over many state operations, including the occupation of the Amazon. Calha Norte is based on the military's fears that if left undeveloped, these border regions with Venezuela, Colombia, the Guyanas, and Suriname would become rife with drug trafficking, foreign left-wing agitators, and Indian secessionist movements.

In the last four years, more than two dozen military bases and airstrips and roads have been constructed. The people most affected have been the Indians. Despite laws prohibiting mining on Indian lands, the military allowed 40,000 gold panners to penetrate the region using Calha Norte's landing strips,

Indigenous people have the greatest stake in the long-term protection of the Amazon. But their methods of preserving it are not based on the principles of keeping it off limits but on using it without destroying it. According to Ailton Krenak, national coordinator of Brazil's Union of Indigenous Nations (UNI), "Our people use the rainforest our whole lives as a place where there are sacred sites, where there are materials to build houses, where you gather materials to make baskets, make bows and arrows, and everything we need. Indian people don't preserve the forest, they respect it, and live together with it."

A cruel irony of colonization is that Indians must prove they are entitled to land they have lived on for thousands of years. "We didn't come here from outer space," says José Correia of the Jaminaua Indians of Brazil. "We have always lived on this land though we don't have anything to show that we own it."

But it is not only cross-cultural misunderstanding that threatens Indian lands. The realization that some of the world's richest mineral reserves and some of the planet's last stands of hardwoods are on Indian lands has led government, military, and development interests to try to remove indigenous peoples from their homelands.

The native people of the Amazon have confronted these pressures by uniting, on the regional, national, and international level to fight for the strengthening of their legal rights. In the process, they have emerged as the most vital component of the international movement to save the rainforests.

The Coordinating Body for the Indigenous Peoples' Organizations of the Amazon Basin states, "Our accumulated ▶

exposing the Yanomami Indians to new deadly illnesses and polluting the rivers that Indians depend on for survival.

In the mid-1980s, native rights groups, such as the Commission for the Creation of a Yanomami Park (CCPY), began lobbying for protection of Yanomami rights and publicizing the Yanomami's alarming health problems that have left 1,500 to 2,000 dead.

In August 1987, the government responded by expelling CCPY, church workers, and other Yanomami support groups from the area and a year later reduced the legally recognized Yanomami territory from 8 million hectares to less than 2.5 million. Furthermore, the protected lands are not even grouped together, but broken up into parcels and separated by national forests that are open to mining and other development.

The plight of the Yanomami does not bode well for the 40 other tribes living within Calha Norte boundaries. Like the Yanomami region, those Indian lands sit atop valuable mineral deposits: gold, "strategic minerals" (niobium, vanadium, and beryllium), cassiterite, diamonds, uranium, thorium, manganese, and petroleum. Brazil has already extended preference-right leases on Indian lands to dozens of domestic and foreign companies including Anglo-American/Bozano Simonsen (South Africa/France), and Brascan/British Petroleum (Brazil/Canada/U.K.).

As big as Calha Norte seems, it is still only one part of the entire security plan encircling Brazil's Amazon. The other part, which encompasses the remaining 3,200-kilometer border region of the southern Amazon with Colombia, Peru, and Bolivia, is called the Program of Development for the Western Amazon Borders (PROFFAO). Under PROFFAO, Indians of the southern border region will suffer the same fate as the indigenous groups in Calhe Norte.

The Petroleum Industry: Drilling into the Rainforest

Like commercial logging, most oil and gas development in the Amazon takes place within Indian territories. Its operations initially destroy and pollute the immediate rainforest areas, but soon, company roads open the entire region to colonization before the Indians can legally protect themselves and their lands.

Oil and gas development is concentrated in Peru and, especially, Ecuador, where the government has made oil development the top national priority. Since oil was discovered in Ecuador's Amazon in 1967, it has became the nation's major source of income. And a 1988 law ensures that subsurface rights in all areas —including Indian reserves and other "protected" areas—belong to the government and are subject to exploitation.

The statute spares no region, not even the internationally reknowned Yasuni Biosphere Reserve. Containing some 679,730 hectares, Yasuni is the biggest natural reserve in Ecuador and one of the largest in South America. Its extraordinary plant diversity includes 4,000 to 5,000 flowering plants, 500 species of fish, 600 species of birds, and 120 species of mammals—many unique to this area.

Under the present oil and gas plan, Yasuni and Cuyabeno parks fall within the 4.5 million hectares being divided into more than 22 blocks, or concessions, for exploitation by the Ecuadorian state (Petro Ecuador) and foreign corporations. The most controversial concession, block 16, is leased to Conoco (U.S.) and located entirely within Yasuni. In 1990, Conoco plans to build a 70-kilometer road straight through the biosphere to link its wells with the cross-country oil pipeline.

At greatest risk are the Huaorani Indians, whom the government granted a protectorate 160 kilometers southwest of Yasuni. But the Huaorani inhabit the entire area from the protectorate through to Yasuni and also consist of bands that are still uncontacted and want to remain so. One band living in Yasuni made this clear four years ago when it killed two missionaries who entered the territory by helicopter to negotiate for the oil companies.

Above: The Quichua Indians of the Huataraco community blockade a road in protest of Texaco and Petro Ecuador oil and gas operations.
Left: Large-scale pollution in the Cuyabeno Fauna Reserve by City's oil and gas operations.
Photos: Gustavo González

knowledge about the ecology of our forest home, our models for living within the Amazon biosphere, our reverence and respect for the tropical forest and its other inhabitants both plant and animal, are the keys to guarantee the future of the Amazon Basin—a guarantee not only for our peoples, but also for all humanity."

Since the late 1960s, Indian organizations have steadily grown in number, size, and effectiveness. Most are organized by federation. Structured to keep decision making decentralized and democratic, federations are made up of delegates from each community, who set an agenda and elect directors to act on their behalf. Often the federations join regional and national confederations to organize more effective political campaigns.

By the 1980s, a number of Indian organizations—including the Union of Indigenous Nations of Brazil (UNI), the Inter-Ethnic Association for the Development of the Peruvian Forest, the Confederation of Indian Nationalities of the Ecuadorian Amazon, and the Central Organization of Indigenous Peoples of Eastern Bolivia—had joined forces in a regional alliance known as the Coordinating Body for the Indigenous Peoples' Organizations of the Amazon Basin (COICA). One of COICA's priorities has been to strengthen relations between the native groups and the international rainforest movement by delivering simple advice to environmentalists: "Work directly with our organizations on all programs and campaigns that affect our homelands."

With title to their traditional lands, Indian communities can then decide what form of community development—if any—they want, based upon ethnic, environmental, and economic considerations. COICA's president, Evaristo Nugkuag, calls this "ethnodevelopment," which can succeed only if the community retains absolute control over development projects and over the production and distribution of goods. Ethnodevelopment projects, says Nugkaug, should also strengthen and restore cultural values and ecological balance devastated by colonization.

In Ecuador, the Organization of Indigenous Nations of Pastaza (OPIP) started one of the first ethnodevelopment projects in the Amazon. OPIP's wood products center in El Puyo is completely run by a Quichua community, from selective logging and reforestation to processing and ▶

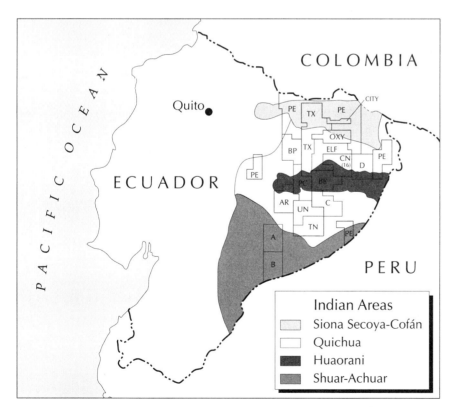

Corporations holding oil and gas concessions: PE = Petro Ecuador, TX = (U.S.), OXY = Occidental (U.S.), ELF = Elf-Aquitaine (France), CN = Conoco (U.S.), BP = British Petroleum, PC = Petro Canada, AR = Arco (U.S.), UN = Unicol (U.S.), BR = Bras Petro (Brazil), TN = Tenneco (U.S.), City (U.S.). A, B, C, D are unleased.

The oil companies claim that most deforestation is caused not by their operations but by colonists who use the company roads to carve out homesteads in the rainforest. This was the case in Petro Ecuador and City (U.S.) oil concessions in the northern blocks where thousands of settlers flooded into the Cuyabeno reserve and Quichua Indian lands. It is the government's responsibility, company officials claim, to stop this situation from recurring in other concessions.

After investigating the Quichua areas in 1989, Judith Kimmerling, a researcher with the Natural Resources Defense Council (U.S.), determined that the direct damage created by the corporations is more extensive than they admit:

• Exploration requires the clearing of thousands of hectares of rainforest for helicopter landing sites, drilling sites, and seismic investigations (involving numerous underground explosions) and has resulted in widespread erosion, sedimentation and contamination of streams, rivers, and lakes, and destruction of fisheries, game, and agriculture.

• Drilling produces toxic wastes — petroleum, gas, bactericides, thickeners, and anticorrosives—that are discharged on land and into streams. When the well is in operation, it burns off 95 percent of the extracted gas and dumps highly toxic wastewater into nearby pits, which kills wildlife instantly and eventually ends up in the river system.

• Oil spills are so frequent that in some areas, streams often burst into flames. Oil is pumped 400 kilometers over the Andes to the coast through a pipeline that can rupture for days before it is stopped. Since it was built 20 years ago, 30 major spills have been reported along the main section alone, with a loss of 63.6 million liters.

Making matters worse, oil and gas development will not sustain Ecuador's economy in the long run. Initially, proceeds (usually in the form of oil or gas) from foreign-operated concessions are divided between the company and Petro Ecuador. Eventually Petro Ecuador reimburses all the foreign company's expenses and after about 20 years takes over the entire operation.

As a result, foreign companies tend to build facilities and extract petroleum as soon as possible, thereby increasing the chances that Petro Ecuador will be left with low recoverable reserves and facilities that are old and contaminated.

Timber Industry Rips into the Rainforest

Commercial logging in the Amazon is now directly responsible for one-fifth of all rainforest destruction. In the last few years, lumber companies have clawed their way deeper into the Amazonian rainforest, penetrating Indian reserves to get at the remaining hardwoods such as mahogany and rosewood, which fetch top dollar in the U.S. and Europe.

Most logging operations in the Amazon do not clearcut but selectively extract the oldest of the 20 or 30 most valuable species. Nevertheless, the surrounding areas are permanently scarred by the falling timber and the dragging of the immense logs by heavy machinery. This technique also robs the ecosystem of the best specimens of the species, ruining chances for their regeneration. But worst of all is that the logging roads open up the rainforest to colonization and other development.

Timber critics claim that even the most well-intentioned attempts by officials, development banks, and environmentalists to limit the areas for logging and institute reforestation are doomed to fail in the face of the government ineptitude and corruption, corporate treachery, and racism that pervade the frontier regions.

Case at point: the first debt-for-nature swap in Bolivia. The highly touted deal in 1987 was intended to create a 10.4 million hectare buffer zone of controlled development around the 540,000 hectare Beni Biosphere Reserve. This meant that timber companies were supposed to replant extracted species. But in 1989, Indians and other observers revealed that since the swap, some 60 percent of the buffer zones' hardwoods have been extracted, mainly for export to the U.S., yet not a single sapling had been planted.

The highest rate of deforestation in Brazil took place during the same period, which included the felling of some 33 percent of Brazil's mahogany, mainly from Indian reserves in Rondônia. According to a Brazilian Parliamentary Commission inquiry into the Rondônia situation from September 1987 to April 1988, this

manufacture of furniture. Other examples are the Yanesha Forestry Cooperative in Peru, which markets hardwoods abroad, and food and handicraft cooperatives of Venezuela's Indians.

These resource management projects are invaluable models for native groups struggling to balance the old ways with the modern necessity for manufactured goods. But for such projects to become a widespread reality, native leaders say, the education system must teach each new generation respect for traditional values while at the same time equipping them with the skills necessary for ethnodevelopment.

Native people, then, must wrest control of education systems from the grip of the missionaries. This is no small task in the Amazon, where trained native teachers are scarce and missionary groups have long held government contracts for education programs throughout the region.

In the Upper Ucayali region of Peru, Shipibo bilingual teachers formed the Federation of Native Communities of the Upper Ucayali and the United Syndicate for the Workers in Bilingual Education of the Ucayali to organize the community struggle for alternative education. "It's just a small step in the right direction," says Cecilio Soria, a Shipibo organizer and one of the first Peruvian Indians to attend the university in Lima.

In Ecuador, the Shuar figured out a unique way to overcome the shortage of native bilingual teachers. Every morning, teachers' aides in hundreds of Shuar primary and secondary schools interpret the lessons of *telemaestros* who broadcast the day's lessons, traditional music and folk tales from the Shuar Federation radio station in Sucúa. After school the children return home to learn the ways of forest life.

Similar efforts to organize Indian-run medical projects are also underway. In Peru, Evaristo Nugkuag reports that the Aguaruna and Huambisa Council (CAH) maintains its own health programs in communities throughout a 200,000-sq.-km. region, providing both traditional and Western health care and training health workers. CAH also provides training and facilities for the production of snakebite anti-serums.

One of the most promising native development projects is now underway in Goiânia, Brazil, where UNI is establishing the Center for Indian Research and Training on Resource Management. The center, the first of its kind anywhere in the world, will not only train indigenous people to manage natural resources, but will also work to restore the damaged rainforests and replenish endangered species.

With such programs starting to emerge, native peoples need allies in the scientific, technical, and foundation worlds. But just as importantly, they need allies to secure control of the traditional lands on which their culture and vision of the future is based.

Indigenous Amazonians have taken great strides toward this goal in the last few years. In 1988, UNI joined with the rubber tappers and other forest dwellers to form the Alliance of the Peoples of the Forest in order to work together on long-range projects.

That same year, Amazonian Indians strengthened ties with environmentalists when the Kayapó and other groups of the Xingu River region organized an international conference in Altamira to oppose a series of dams. In 1989, a COICA delegation followed with a tour of the U.S., offering a framework for a more productive dialogue with environmental groups. Its foundation is simple: Indian nations in the Amazon not only have thousands of years of history and culture, but also a vision for the future, where the survival of their people and the protection of the forest are one. ◆

Traditional slash-and-burn agriculture of the indigenous people.
Photo: ©Aguirre/Switkes

extraction was illegal, the Indians who had signed lumber contracts were scandalously cheated, and widespread complicity had taken place between the lumber concerns, national Indian agency officials, and the Brazilian Institute for Forestry Development. As usual, there was no legal action in the matter.

It's hard to tell who benefited most from this deal: government officials, lumbermen, or the U.S. The biggest U.S. tropical hardwood import is mahogany. In 1988 Brazilian supplies peaked to almost two-thirds of U.S. mahogany imports — three times the 1978 volume. That same year, Brazil also became the top supplier of tropical hardwood veneer to the U.S., providing 15 percent of the total.

Indians were clearly the losers even when they refused to sign lumber contracts. Lumber concerns operate on Indian lands with or without permission and will go to any length to uphold that prerogative. In March 1988, after local lumbermen were evicted from recently demarcated Tikuna Indian lands in Brazil, they opened fire on and hunted down unarmed Tikuna, killing 14 men, women and children and wounding 23 others. In another well-documented case in Peru, lumber concerns have forced Ashaninka Indians into virtual slavery to extract wood.

A growing number of Amazonian Indians are now demanding a key role in land management that affects them and are becoming more adept at exposing the failures of management schemes imposed from the government, conservation groups, and development banks.

Forest Gatherers

The ranchers who ordered the murder of Chico Mendes in December 1988 silenced one of Brazil's most eloquent and popular defenders of the rainforest. But it will not be so easy to slow the cause Mendes died for: the fight for land rights and economic justice for traditional forest workers.

Francisco "Chico" Mendes Filho has often been portrayed in the international media as an ecologist, but he was above all a tireless organizer for Acre's Union of Rural Workers. Mendes was a rubber tapper, or *seringueiro*, one of hundreds of thousands of riverine people and forest dwellers, including fishermen, nut gatherers, and others who earn a living in the Amazon by selling products they gather from nature. Mendes' mission was to unite forest gatherers in a struggle for justice and a better life for a people who have been one of the most exploited groups in the region.

The economic history of the Amazon has been marked by booms and busts, caused by fluctuations in the international demand for forest products. From the *drogas do sertão*—spices, dyes, perfumes, and fruits—valued by the European aristocracy to quinine, chicle, tiger hides, and more recently, gold, bauxite, and coca, the story of Amazonian development has been in response to the demand for export goods.

In the late nineteenth century, the emergence of bicycles and automobiles created the need for rubber tires, thrusting the Amazon into prominence as the world's main

Wasting the Rainforest with Cattle Ranching

Cattle ranching is the greatest cause of rainforest destruction in the Amazon Basin and is responsible for more deforestation than all other systems of production combined. Amazonian governments, however, continue to support it despite well-documented evidence that it is the most environmentally and economically unsound form of land use ever employed in the rainforest.

In the Brazilian Amazon, livestock is the principal land use, accounting for more than 85 percent of the total rainforest cleared to date. Small farmers play a role in initiating the forest-to-pasture conversion process. They find that clearing the rainforest with fire leaves behind ash deposits that increase nutrients in the infertile soils—but only for about three years. Then the farmers have to abandon the land and start over again somewhere else or sell out to ranchers.

Ranchers themselves only have about a five-to-seven year period of grazing before soil nutrients necessary for pasture growth are depleted by erosion and leaching. Pastures are degraded even further by soil compaction and invasions by noxious weeds inedible to cattle. Pastures become so degraded that more than 50 percent of the cleared areas have been left unused.

Nevertheless, cattle ranching is still one of the most profitable investments in the Amazon. Through generous subsidies, tax holidays, and other fiscal incentives the government makes it almost impossible not to turn a profit from a large cattle ranch. Ranchers clear the forest, sell the wood, and plant cattle pasture to establish "proof of occupation," which legitimizes land claims. Once this is done, the land owner may not even bother to raise cattle. With the runaway inflation of the

Amazonian countries, land speculation is one of the smartest investments.

When ranching takes place, it is a low-cost operation because it only takes one ranch hand to manage almost 1,000 hectares of pasture. In addition, government-sponsored road projects make it easy to market the cattle and timber. These roads have increased more than 13 percent a year in Amazonia—from 11,802 kilometers in 1960 to 24,584 kilometers by 1984.

By promoting speculation, the government ignores the fact that most of the lands in question are usually occupied by peasants, rubber-tappers, or indigenous people. When there is a land dispute, the authorities usually stay out of it, leaving the ranchers to take matters into their own hands.

This policy has resulted in approximately 540 assassinations in the Amazon between 1985 and 1988. Most of the killings have been attributed to cattle ranchers affiliated with Rural Democratic Union (UDR), a national right-wing organization of landowners. UDR's formed in the mid 1980s to lobby against effective land-reform measures being included in the new constitution—a goal it most definitely attained.

But fearing the rise of popular agrarian reform movements by the landless, rubber tappers, and Indians, the UDR has stockpiled arms and formed paramilitary groups, which make good on the death threats routinely delivered to priests, elected officials, judges, and union organizers such as Chico Mendes.

The violence and vast environmental destruction from cattle ranching has sparked innovative research into the most productive uses of Amazonian rainforest. Not surprisingly, ranching productivity figures are the worst and sustainable uses are the best. According to Charles Peters at the New York Botanical Garden, the net revenue generated by sustained harvesting of marketable fruits, nuts, latex, and occasionally logged timber in one hectare of primary Peruvian Amazon for 50 years is worth $6,330 if sold in local markets, and only $2,960 if converted to cattle pastures. Each hectare can support only one cow.

Brazilian seringueiro. Photos: ©Aguirre/ Switkes

supplier of latex. During the rubber boom, rubber barons *(seringalistas)* were granted concessions on huge tracts of land. In Bolivia, the Suarez brothers alone controlled 62,160 sq. kilometers of productive forests. However, governments and rubber entrepreneurs hit a formidable obstacle: Wild rubber trees were scattered through the forest and had to be tapped.

The *seringalistas* faced a chronic labor shortage, largely due to the genocide carried out against indigenous peoples. Thousands more were enslaved and killed as the fever for rubber took hold. To guarantee a labor supply, peasants from economically depressed areas such as northeastern Brazil were lured to the Amazon with the promise of quick fortunes.

The rubber tappers' dreams of prosperity soon turned into nightmares. In the *seringais*, isolated encampments deep in the forest, their lives were tightly controlled by their bosses. The tappers were locked into a system of debt peonage, wherein traders sold the only available supplies of food, tools, kerosene, lamps, and other essentials at inflated cost—in exchange for rubber, for which the traders paid unfairly low prices. Tappers who questioned their authority were tortured or killed. Valdemar Barbosa, a rubber tapper in Acre, recalls: "In the old days, the bosses murdered rubber tappers to not have to pay salaries. … They burned the rubber tappers with hot wax, and then took them out into the forest and killed them."

As growers smuggled rubber seeds to Malaysia, where rubber could be grown on plantations, the demand for ▶

Gold mining in Rondônia, Brazil. Above: Dredges filter sediment through mercury-laden sieves. Left: Gold is purified by heating the amalgam and evaporating the mercury. Photos: ©Aguirre/Switkes

Amazon forest rubber subsided, and the region's economy fell into rapid decline. Many *seringalistas* abandoned their estates, moving into more profitable activities. The now-autonomous *seringueiros* of Acre became the heart of the movement led by Chico Mendes.

In the 1960s and 1970s, the opening of highways into Acre and a boom in government-subsidized cattle ranching represented a new and deadly threat to Acre's *seringueiros*. A large number of rubber tappers were evicted by cattle ranchers and were forced into the slums of cities like Rio Branco, but many were determined to stay.

Rubber tappers developed a strategy of direct, non-violent action—*empates*, or standoffs—to defend the forests. Often, entire communties, including women and children, arrived to confront work crews clearcutting forests for cattle pasture.

By the late 1970s, the *empates* were making ranchers extremely nervous, and they responded with violent attacks on leaders of the rubber tappers' union. Chico Mendes realized the workers' movement stood little chance of responding in kind to such violence. He worked to broaden the *seringueiros'* movement by initiating literacy campaigns and organizing cooperatives, allowing rubber tappers to process and directly market their rubber and Brazil nuts as a path to economic freedom.

Mercury Pollution: The True Price of Goldmining

More than 400,000 miners struck with gold fever have descended on the Amazon. In hopes of striking it rich, they are using a gold extraction process so hazardous, scientists say, that it is creating a major disaster of intoxication and environmental contamination—affecting not just the mining sites but the entire river system and the downstream populations.

Mercury, the metal used to extract the ore, is one of the deadliest toxins on earth. After the river sediments are dredged from the river bed, they are filtered through mercury-laden sieves to separate and amalgamate the gold. Then the gold is purified by heating the amalgam and evaporating the mercury. It takes two to ten tons of mercury to purify one ton of gold, and as much as 1,000 tons of mercury are being dumped or burned in the Amazon each year.

Exposure to small amounts of mercury on a daily basis—whether inhaled, absorbed through the skin, or ingested—can cause severe poisoning. Mercury penetrates and accumulates in the central nervous system, at first causing irritability, fever, depression, and problems with hearing, kidneys, sleeping, and memory. In more advanced stages, it causes madness, birth defects, and death.

Due to a lack of funds, trained personnel, and commitment by the governments, systematic health and environmental studies have yet to be conducted in the Amazon watershed. Limited research in Brazil, however, has shown, what many experts fear are just the first ticks of the biological time bomb:

• Miners are at greatest risk because they handle and breathe mercury continuously, use eating utensils that contain mercury particles, and consume contaminated fish. One preliminary study of miners' hair and urine showed mercury levels up to 20 times the recommended limit by the World Health Organization (WHO).

• A veterinary pathologist has reported unusual cases in gold mining areas of spastic calves, deformed pig fetuses, and even some children with birth defects similar to those with Minimata disease in Japan.

• **The Kayapó Indians are just one of numerous Indian tribes and riverine populations living downstream from mining activities. Health studies several years ago showed 25 percent of the Kayapó tested contained concentrations of mercury up to 14 times the level considered safe by WHO.**

• **A study by the University of Rio de Janeiro found fish in the Madeira River contained five times the permitted mercury levels for consumption. Another study found fisheries, sediments, and plants contaminated up to as far as 1,600 kilometers downstream from those mining operations.**

The health risks could substantially worsen if Brazil goes ahead with its plans to build 70 high dams in the Amazon. Most of these dams, located downstream from mining areas, will have reservoirs of up to 2,300 sq. kilometers that contain the mercury-laden sediments, creating giant contaminated lakes throughout the region.

Scientists, physicians, and anthropologists who gathered for the seminar "Control of Mercury Contamination in the Amazon" in Belem, Brazil, in December 1989 noted that there are still no alternatives to mercury use in gold mining. Mercury is the easiest way to recover the small amounts of gold in the common alluvial deposits. Methods could be devised to reduce inhalation of the mercury vapors, but not its release into the environment.

Governmental restrictions on the importation of mercury, such as those in Brazil, are useless, participants concluded, because they are unenforceable. The miners smuggle mercury supplies into the country just as easily as they smuggle the gold out to avoid paying taxes.

In 1985, more than 100 forest gatherers from throughout the Brazilian Amazon gathered for the first time to create the National Council of Rubber Tappers. The meeting produced a series of demands for economic and social reform: an end to debt peonage; better health care and education; credit and price guarantees; applied research on alternative forest products such as fruits, nuts, and medicines; and the suspension of tax breaks for cattle ranching and logging activities. A more recent meeting included Brazil nut gatherers from the northern Amazon, and representatives of gatherers' associations in Bolivia.

From these meetings emerged a grassroots proposal for sustainable development in the Amazon—the concept of the extractive reserve. The proposal calls for the granting of use concessions to local populations who depend upon on the management of forest resources for their survival. Within extractive reserves, forest clearing would be limited to small areas, primarily for subsistence agriculture.

"We understand that Amazonia cannot be transformed into an untouchable sanctuary," Mendes observed. "On the other hand, we also understand that there is an urgent necessity to avoid deforestation that is threatening the Amazon and the life of all people on this planet."

The concept of extractive reserves could also be extended to other populations whose survival depends upon the sustainable use of natural resources, such as riverine peoples and fishermen. Throughout the Amazon region, dozens of forest products, including jute, chicle, palm hearts, tree fibers, fruits, ▶

Delia Marlene's riverine community in Ecuador is at risk if gold mining spreads throughout the Amazon. Photo: Nicki Irvine

and nuts are gathered by hand from the forest. Most of these workers still live under the yoke of debt servitude borne by rubber tappers.

Raimundo de Barros, a union leader in Acre, said forest people are not interested in owning large areas of land. "What's important to us," he said, "are the streams that are within the area that provide fish, and the forest itself that has Brazil nut trees, *bacaba* and *açaí* fruits, and rubber. ...We are human beings. ...We've lived our lives enslaved and exploited, and we feel this has to change."

Fishermen, too, find their livelihood threatened as commercial fishing enterprises arrive, using destructive techniques such as explosives to stun or kill fish, and purse seines, which indiscriminately net whatever falls in their path.

Along the upper Amazon (Solimões) of Brazil, tensions between fishing communities and commercial fishing boats reached a crisis in 1987. Members of a local fishing colony, seeking to protect traditional fishing grounds in a lake above the floodplain, confiscated the fishing gear of several commercial boats. Riverine communities along the Solimões have recently formed committees to maintain an around-the-clock vigil over 34 lakes.

The expanding network of massive hydroelectric dams is also cause for concern among the Amazon's river dwellers. Tucuruí Dam on the Tocantins and Balbina on the Uatumã River have radically changed the nature of the river systems.

Six Arguments Against Hydroelectric Dams in the Amazon

In Brazil, the construction of more than 70 hydroelectric dams within the Amazon river system in the next 20 years has been a crucial part of the Energy Sector's Plan 2010. Cheap available energy is what planners hoped would attract investments in the exploitation of Amazonian natural resources, which would in turn generate the funds needed to service Brazil's huge foreign debt. But the construction of the Amazon's first three large-scale hydroelectric dams—Tucuruí, Balbina, and Samuel—show that mega-dams are not only a disaster for the people living along the dammed rivers but also for the entire national economy:

• Dam projects plunge Brazil into deeper debt. Project costs are always more than the original estimates because of construction delays, technical problems, and inflation. Projected to cost $383 million, Balbina ended up costing $750 million in foreign loans to produce 100 megawatts.

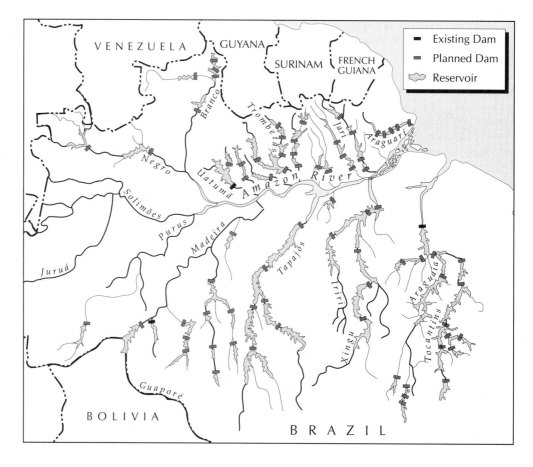

• Electricity from Amazonian dams is the most expensive in Brazil. But it is not the industries who pay going rates; they have been receiving government subsidies as incentives to invest. The burden of covering the higher energy prices and industrial subsidies falls on the average ratepayer and taxpayer in Brazil.

• Amazonian dams have a short lifespan. In all three dams, flooding of rainforest creates a massive swamp of decaying flora and fauna. This process raises the acidity of the reservoir water that corrodes the dams' turbines and shortens their lifespan from the average of 30 years to 20 or even ten.

• Dams built in the Brazilian Amazon destroy more rainforest than anywhere else in the world. If all planned Amazonian dams are built, more than 25,900,000 sq. kilometers of rainforest— an area about the size of the United Kingdom—will be drowned. The low-lying Amazon basin requires that a relatively huge area be flooded to fill a reservoir. For Balbina, more than 2,300 sq. kilometers had to be flooded to fill the reservoir to its total average of only 7.4 meters. Drought, however, delayed filling to even this level for months. At Samuel, engineers had to build 48 kilometers of dikes to help create a lake of 458 sq. kilometers.

• Displaced populations are never compensated. After losing their farms and ancestral homelands to the dam projects, many subsistence farmers from Tucuruí were forced to move to shanty-towns to find work, while tribal groups, such as the 400 Waimiri-Atoari from the Balbina area, were forcibly split up and relocated into areas where they have been exposed to deadly diseases and cannot sustain a livelihood. Some 500,000 people will suffer similar fates if all planned Amazonian dams are built.

• Pollution caused by dams is a major health hazard and environmental catastrophe. Tucuruí reservoir's immense body of stagnant water has turned the area into a malaria-infested region. The decaying fauna and flora of the Balbina dam has poisoned the reservoir, the downstream fisheries, and freshwater supplies of the riverine communities for hundreds of kilometers.

Riverine family in Peru. Photo: Alice Levey

A massive fish kill after the closing of the floodgates was followed by bacterial contamination of the river water, rendering it unfit for drinking or bathing purposes. Riverine people say that fish now spoil rapidly, and that their taste and color have changed. Skin irritations are more common, caused by swarms of insects once unknown to the area.

Committees of the Victims of Tucuruí and a Committee for the Defense of the Uatumã have been formed, and they have participated in national Encounters of Peoples Impacted by Dams. The riverine peoples have pressed for a review of the government energy policies which call for the construction of dozens of new hydroelectric dams, and are demanding adequate provisions for relocation and compensation for those affected by already existing projects.

Another focus of riverine peoples has been to protest against the poisoning of rivers with mercury from gold-mining activities. Riverine populations have demanded government action to limit miners to specified areas, and to require techniques to prevent mercury from entering into aquatic ecosystems.

Forest gatherers are only beginning to become organized as a class, and feudal dominance by river traders still remains the rule in most parts of the Amazon. But their activism in defense of the forest has inspired a fresh look at the role of forest peoples in promoting sustainable development. ◆

The Colonists

In the past two decades, waves of migrant farmers, landless rural workers, and gold miners have become the newest population group in the rainforest. Driven from other regions of South America by land conflicts, the expansion of agribusiness, and political violence, and drawn to the Amazon with the promise of free and abundant land and prosperity, they have become the latest victims of the chaotic situation on the frontier.

Colonization projects in the Amazon were designed as a way to avoid land reform, serving as a pressure valve to relieve social tensions in other regions such as in the south and northeast of Brazil, and in the Andean highlands. With a few landlords controlling vast estates—in Brazil, less than 1 percent of land owners control nearly half of the country's arable land, and in Colombia, one-tenth of landlords own 80 percent of the farmland—small farmers are given little choice but to blaze a trail in the rainforest.

These colonization schemes, including the Transamazon Highway and the Polonoroeste Project in Brazil and Pichis-Palcazu in Peru, have promised to provide land for millions of landless peasants, but have resulted instead in widespread disillusionment. More than 1 million migrants arrived in Rondônia, Brazil, during the 1980s, and 15,000 came to San Martin Province in the eastern slope forests of the Peruvian Amazon between 1973 and 1983.

Infertile soils and a lack of roads and agricultural assistance and credit have made farming difficult, even at a subsistence level. Malaria and other debilitating diseases at epidemic levels have left the *colonos* and their families unable to perform the arduous clearing and planting required to survive on the frontier. Many of these farmers have ended up selling or abandoning their land and pushing further on into the forest, or to the periphery of large cities, where they live under marginal conditions.

The new settlements have also meant the clearing of vast areas of the forest. In Rondônia, one-quarter of the rainforest has been cleared. Sixty-two thousand square kilometers of Peru's forests—seven percent of the Peruvian Amazon—has also been deforested. Similar problems have emerged in other Amazon nations. In Ecuador, land-seeking migrants have followed roads plowed by companies searching for oil, leading to clashes with indigenous forest ▶

Colonos family in the Arara Indian Reserve, Brazil.
Photo: ©Aguirre/Switkes

Satellite images show the progression of land clearings in the same area of Rondônia from 1976-1981.
Photos: Woods Hole Research Center

Polonoroeste: A Model of Integrated Regional Destruction

In 1981, Brazilian and World Bank leaders decided to open up the Amazon for the landless by creating the Polonoroeste resettlment scheme. Then, it was hailed by its planners as a promising model of "integrated regional development." But within just a few years, Polonoroeste had plunged hundreds of thousands of migrants into misery and was causing the highest rate of deforestation in the world.

At first, Brazil did not intend to create a project the magnitude of Polonoroeste. The original plan was to asphalt the BR-364 jungle highway on the country's northwest frontier, linking Cuiabá and Porto Velho, the capital cities of Mato Grosso and Rondônia. But due to pressures from the World Bank, the highway project was expanded to ensure the "orderly socio-economic development" of the region influenced by the road.

Within an area three-quarters the size of France, Polonoroeste was to ensure the demarcation of tribal lands and the provision of social services for Indian communities, protection of biological and other forest reserves, and the promotion of sustainable agriculture among farmers. By the mid-1980s, Polonoroeste was in chaos:

• More than 2 million hectares of tropical forests in Rondônia were cleared or torched. Virtually every indigenous and forest reserve in Rondônia had been invaded by colonists, loggers, and speculators, often resulting in violent conflicts and the expulsion of original inhabitants.

• Following the asphalting of BR-364 in late 1984, migration rates to Rondônia more than tripled, reaching over 150,000 newcomers in 1986 alone. This unpredicted migratory avalanche was due not only to easier travel, but also to soaring poverty elsewhere in the country, and the government's propaganda hailing Rondônia as Brazil's new "El Dorado."

• Northwest land values and deforestation rates shot up soon after the infrastructure projects were complete and frontier development policies enacted. By law, converting forest to cattle pasture or farms is an "improvement" to land and establishes a land claim. Such policies encouraged both rich and poor migrants to indiscriminately clear forests to obtain land titles.

• Nearly 80 percent of the migrants could not afford to keep their farms because they contained infertile soils. Project planners recommended monocropping for export agriculture, which relies heavily on expensive labor, fertilizers, and pesticides. Meanwhile, crop prices were dropping due to fluctuations on the world market, exploitation by local merchants, and soaring interest rates for agricultural credits (created by the IMF's austerity measures). Rather than go into debt, most of these people abandoned their lands to ranchers.

• The government ignored the horrific problems and continued to construct roads, initiate settlement projects, and grant land concessions in areas that were often occupied by forest dwellers.

Despite the growing international criticism of the Polonoroeste project, the World Bank refused to cut off project funding, except for several months in 1985 to force Brazil to demarcate land for the Uru-Eu-Wau-Wau Indians, and to remove squatters from a few other Indian and forest reserves.

Even though the Brazil government has recently cut the Uru-Eu-Wau-Wau Reserve to a small fraction and ignored most other environmental and social conditions of the previous loan, the Bank now considering a new Polonoroeste loan for $200 million.

The Bank claims that new conditions promoting "agro-ecological zoning" in Rondônia will somehow limit homesteading, ranching, logging, and agriculture to specific zones—once again, without any realistic way to enforce such rules.

The Highs and Lows of Growing Coca

Hundreds of thousands of farmers have been migrating to the Amazon each year to grow coca, the single most lucrative commodity in Latin America. The growing demand for cocaine in the U.S. and Europe has spawned this vast illegal monoculture, which is destroying the very rainforest it counts on to shield it from the law.

It is legal to grow a limited amount of coca in Peru, Bolivia, and Colombia because of its cultural and practical uses. It has been chewed for centuries by Indians of the Andes and upper valleys of the Amazon for religious rituals, for strength to farm, to adapt to high altitudes, and to alleviate hunger. In Bolivia, some 40 percent of all households use coca.

But in the last 15 years a combination of factors—cocaine's growing demand abroad, soaring domestic poverty, and the emergence of the international marketing network run by the drug mafia—has convinced hundreds of thousands of farmers in Peru and Bolivia to penetrate deeper into the rainforest to grow illegal coca. After harvest, they process it into paste, and ship it to Colombia where it is refined into powder and exported, mainly to the U.S.

Today, official estimates put the number of small farmers involved in coca roughly at 500,000 and cocaine shipments from the Amazon at more than 400 tons a year. Farmers now make $720 per load of coca (135 kilograms) and wage earners can make $12 per day—five times the average pay.

The national economies of these countries now depend on the income from cocaine more than any other industry. Coca production yields are estimated at $3 billion—roughly equal to half of Peru's annual legal exports. Conservative estimates put investments into the Colombian national economy from the drug cartels at $1 billion per year. They could easily be five times as much.

Such large investments are carefully guarded. In Peru, coca operations are protected and brokered by the Maoist guerrilla group Sendero Luminoso, which earns $30 million for its services. In Bolivia and Colombia, influential families, politicians, and army personnel profit handsomely from playing the same role.

peoples. In Colombia, the ongoing civil war and the emergence of cocaine as a valuable commodity have spurred a rapid migration to the rainforest.

Although these impoverished migrants have often invaded the territories of Indians, rubber tappers, and other forest dwellers in their search for land, they, too, are impacted by rural violence. Peasants are often forcibly evicted by cattle ranchers and land speculators who seek to establish claim to lands the farmers have cleared. In some areas, such as the "Parrot's Beak" of Brazil's eastern Amazon, slave labor, torture, and other forms of brutality have become common occurances. More than 1,000 rural workers have been murdered in the past decade.

According to Raimundo Cordeiro da Cruz, who worked on a ranch in the south of Pará, "If a worker objected, they took him and killed him. One *companheiro* was carried into the forest, told to dig a hole, and they killed him and threw him into it. I found a hole full of the bones of bodies they had buried."

In order to intimidate rural leaders, Brazil's Democratic Rural Union, the political wing of the ranchers, has reportedly stockpiled weapons, and hundreds of leaders in grassroots movements of rural workers have been threatened or killed. The ranchers who most often gain control of these lands, either by buying out the frustrated farmers or by coercion, are likely to burn and clear even larger areas, in order to prove their "productive use" of these areas for purposes of establishing a land claim.

In the western Amazon, the principal cash crop for small farmers is the coca leaf. Coca, a native medicinal and ceremonial plant, is also the raw material for cocaine. In many ways, coca is a perfect crop for the Amazon, because it grows in poor soils and can be harvested four times a year.

In Peru, coca growers are the targets of a U.S. government-financed program to chemically eradicate coca. In Colombia, coca is the latest in a series of "booms" that has swept the region, and farmers earn far more for growing the illegal leaf than they would for growing manioc or bananas, although this leaves them susceptible to fluctuations in the price the drug mafia is willing to pay for the leaf.

Farmers in Peru and Colombia are caught in the middle of a power struggle between the military, drug lords, and guerilla fighters, which makes survival precarious and long-term, ecologically sound agricultural planning impossible. Many have joined guerrilla groups as a way of trying to take control of their destiny.

The discovery of gold and other valuable minerals, and the prospect of striking it rich, have also fueled new waves of migration to the Amazon. Today there are 600,000 gold miners throughout the region, who are scouring river ▶

The Movement of the Landless helps the farmers of southern Brazil hold on to fertile lands and avoid migration to the Amazon.
Photo: ©Aguirre/Switkes

For all the intrigue of the economic and political implications of cocaine production, little attention has been paid to the environmental impacts. Coca—the "Attila of tropical cultivation"—is to date responsible for the destruction of 1,600,000 hectares of rainforest in the Amazon. About 1 million hectares of this loss are in the Huallaga Valley in Peru, an area responsible for 60 percent of the world's coca, where growing has increased sevenfold since 1975 and is intruding into Cutervo and Abiseo National Parks and the Alexander Von Humboldt and Apurimac National Forests.

The watershed is also being assaulted with huge amounts of chemicals used for processing the coca leaves into a paste. Herbicides, including Agent Orange and Paraquat, are used to treat 5 billion coca bushes. According to a 1987 study, chemicals used and dumped in the Huallaga include 56.8 million liters of kerosene, 30.9 million liters of sulphuric acid, 6 million liters of acetone, 6 million liters of solvent toluene, 16,000 tons of lime, and 2,880 tons of carbide. Many fish, reptiles, and crustaceans have already disappeared from the rivers and streams.

If the Bolivian and Peruvian governments would risk public opposition, the U.S. would spray Spike or tebuthiuron, herbicides that kill woody plants and leave the ground barren for years, causing massive erosion. Spraying, however, is considered the only effective way to destroy such huge areas of the hearty coca plant. In 1988, the U.S. spent $15 million clearing coca by hand—a method that merely brought the bushes back even stronger.

With spraying foiled for now, the U.S. has resorted to tactics from the Vietnam War. It has built a $3 million military fortress, complete with moat, airstrips and hundreds of advisors. Huey helicopters, guided by satellite photos, will search out and destroy coca fields, labs, airstrips, and shipping operations deep in the jungles. Both Peruvian and U.S. officials at the base, however, admit that their operations have had little effect. Coca production will keep doubling or tripling, they say, as long as there is a demand for cocaine.

beds and stripping hillsides in search of precious metals. Mining companies and the military have argued that the gold miners be forcibly replaced by large-scale mining operations, giving rise to the charge that the miners are being used as pawns to open up new areas of the forest to powerful economic interests.

At a demonstration in 1988, several hundred miners in Marabá, Brazil, calling for the government to improve conditions at the Serra Pelada mine, were dispersed by military police and have "disappeared," their fate unknown. Gold miners have formed unions to fight for better working conditions, although they often find themselves in conflict with groups of indigenous people and riverine populations.

In recent years, a wide variety of popularly based organizations has emerged among recent migrants to the Amazon frontier. Within small-farmer settlements, for example, migrant colonists have created organizations to promote direct marketing networks, better access to agricultural credit, improved road conditions, and health and education services.

Movements of landless rural workers have also emerged to defend land claims against cattle ranchers and land speculators, and to fight for agrarian reform. Brazil's Movement of the Landless is strongest in the south of Brazil, an area from which peasants are being expelled and forced into the Amazon. Ivonete Campos Ferreira's family is one of 350 that for 18 months have occupied an estate in Paraná. "We can't wait for the government," she says. "We have to organize, because we know there is enough land here for everyone."

In Bolivia, farmers have campaigned to keep the cultivation of coca legal, for use in tonics or for chewing as a mild stimulant. Peasants from the Central Union Confederaion of Rural Workers of Bolivia in the Chapare region negotiated an agreement with the government for substitution of coca with other crops. The agreement was reached only after union leaders went on a 20-day hunger strike to dramatize their demands. Such agreements will require funding from countries like the United States, which until now has proven more willing to fight the "drug war" than to search for sustainable social solutions to the coca growers' situation.

Today, there is widespread recognition that the flow of population to the Amazon must be slowed. While there are strong pressures for development, small farmers and policy makers alike are acknowledging that the quality of life of the people of the region is linked to the survival of the rainforest.

Manoel Alves de Oliveira, a farmer who came to the Amazon 20 years ago, has learned to plant diverse crops. "The future of Rondônia," he said, "merits the preservation of nature. When we moved here, it rained abundantly, and it was never cold. Now, it seems that with the destruction of the forest, nature is giving its response. It's no longer the same Rondônia as when we first came here." ◆

Carajás: A Mining Metropolis and a Living Hell

Ten years ago, the Brazilian government launched the $62 billion Greater Carajás mining project (GCP) in the eastern Amazon, promising that the massive development scheme would be an answer to Brazil's ailing export economy, unemployment, and landlessness. But when poor migrants poured into the area, they found themselves pitted against powerful land speculators and corporate investors who were plundering the environment with no intention of sharing the benefits.

In 1982, Japan and the World Bank helped launch the GCP by loaning the state mining company Companhia Vale do Rio Doce (CVRD) $804 million for the construction of the Carajás iron ore mine, a 900-kilometer railroad, and a deepwater port at Ponta de Madeira. Although the Bank denies responsibility for development beyond the 160 kilometer radius of the iron mine, it helped CVRD obtain $3.6 billion more in co-financing from the European Economic Community (EEC) and other sources.

The Brazilian government used the funding to transform an area of Para state as large as France and Great Britain combined into an export zone for minerals and timber. Investments receive a ten-year tax holiday, exemption from import taxes, easy access to credit, subsidized water and energy, and free land. By 1989 GCP operations included Tucuruí (an 8,000 MW hydroelectric dam), two aluminum plants, and four pig iron smelters.

Other GCP proposed projects include another 36 iron smelters (22 of which are scheduled for construction in 1990), at least one huge hydro dam, and agribusinesses designed by the Japanese to grow soybeans and oil palms for export, and rice, corn, beans, and manioc for domestic markets.

Of all the GCP industries, the pig iron plants are the most environmentally destructive as they are fueled by 2.3 million to 4 million tons of charcoal annually—fuel made from the surrounding rainforest. The government claims that reforestation for charcoal production will be part of the project but the pig iron smelters are economically viable only if they do not include the expense of reforestation.

According to CVRD staff, businesses gearing up to produce pig iron show no signs of implementing reforestation, nor are agro-forestry techniques sufficiently developed to undertake large-scale reforestation projects over the short term. As a result, environmentalists estimate that GCP will deforest 3,500 sq. kilometers, completely denuding the railroad corridor within the next 20 years.

Despite its ambitious plans, GCP has become Brazil's most tragic example of unplanned development. By 1988, GCP had attracted only one-sixth of its target investments and offered only 10 percent of the 1 million jobs originally promised. Settler disillusionment quickly gave way to levels of depravity and wholesale violence unmatched anywhere in the Amazon:

• Shanty towns are proliferating in the region. Cities such as Marabá and Açailândia have swelled by almost 50 percent with hundreds of thousands of migrants who have come in search of work, as well as those displaced by GCP itself.

• GCP is the area of the highest number of assassinations in Brazil. Unrelenting conflicts between landholders and the landless have spawned such a demand for murder that in some towns, gunmen offer price lists for killing priests, union leaders, city officials or ranchers.

• Legal and illegal timber operations have been largely responsible for deforesting about 10 percent of Pará, Maranhão, and Tocantins. Much of these areas has been replaced with cattle pasture.

• Some 10,000 Indians in 25 areas have suffered land invasions by settlers, loggers, and miners.

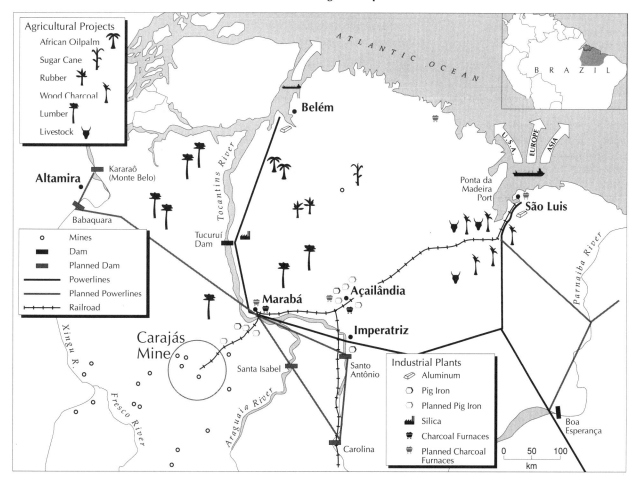

Support Groups

There are more organizations around the world than ever before trying to save the Amazon from destruction. Although most of these groups have their own particular approach to the issue, few can deny that the Amazon will be doomed unless more support is given to the people of Amazonia in their fight to protect the rainforest.

Latin American activists were the first to start tackling the problem during the dictatorships of the 1960s and '70s. These progressive church workers, academics, scientists, human-rights activists, and legal experts recognized that fighting for environmental protection would be to no avail unless agrarian reform, land rights, and democratic constitutions were adopted and protected.

Their strategy was to begin long-term efforts to work for social, political and economic reform by supporting popular movements on the grassroots level. Under great personal risk, the Pastoral Land Commission of Brazil helped organize rural laborers to get better working conditions, protect squatters against violence, and campaign for agrarian reform. Brazilian indigenous people received valuable assistance in their battle for land rights from the Ecumenical Center for Documentation and Information. The group, Aid for the Indigenous People of Eastern Bolivia, provided valuable assistance that led to the formation of CIDOB, the main Indian organization of the Bolivian Amazon. Rubber tappers were supported in their local organizing efforts by the Center for Amazon Workers.

Support groups also worked to make economic alternatives accessible to the people of the Amazon. In Brazil, the Institute of Amazon Studies helped rubber tappers to organize their first national meeting, and to develop the plan for extractive reserves. The Pro-Indian Commission of Acre helped native peoples create cooperatives to market rubber and traditional handicrafts. According to Terri Vale d'Aquino, a founder of the Commision, "Helping an Indian group to keep their land will do more to save the rainforest than one thousand articles in *The New York Times*."

Even before the Amazon was an issue for discussion at the world's breakfast tables, organizations such as Amnesty International, the International Work Group for Indigenous Affairs, and Survival International, were working closely with the Latin American groups to on human rights issues while Oxfam and Cultural Survival played important roles in funding community development projects.

The Good, the Bad, and the MDBs

Amazonian deforestation is inextricably linked to the lending policies of multilateral development banks (MDBs) such as the World Bank and a smaller, regional MDB for Latin America, the InterAmerican Development Bank (IDB).

These MDBs are public lending institutions whose operations and loans of more than $25 billion annually are backed principally by contributions from the governments of the U.S., Japan, U.K., West Germany, and France. The lion's share of contributions, about 19 percent, comes from the U.S.

More than half of the MDB loans in recent years have gone to support development projects in South America that have had the worst records of deforestation and social destruction: export agriculture, rural development schemes such as Polonoroeste and cattle ranching, hydroelectric dams, and road building. Although MDBs do not pay for entire projects, they attract at least twice the sum of their investments in co-financing from the borrower country, other development agencies, and private banks.

In the 1980s, MDBs have gone well beyond their mandate to fund individual projects. Through their more recent structural adjustment loans and sector loans, MDBs now have the power to reorganize entire national economies so that they will be more integrated into the international market and capable of servicing the foreign debt. These loans usually provide subsidies for large land holders to produce export goods and foods that earn hard currency, while austerity measures cut back price supports for social services and basic food items on which the poor depend.

In the 1980s, environmental and human rights advocates reminded MDBs that since their operations are guaranteed with pubic funds, they are therefore subject to public scrutiny and pressure. They demanded that MDBs reform their loan policies by:

• Increasing the number of trained environmental staff in the World Bank from seven to 100 to monitor and influence the hundreds of loans per year.

• Instituting an environmental impact process to review the effects of all MDB-funded projects.

• Making the environmental impact assessment and all other relevant information about projects available to the public for review.

• Involving the affected communities in all phases of project planning and implementation.

Although the MDBs have agreed to meet many of these demands, little has actually been accomplished. The World Bank has created environmental positions, but they still have fewer than 50—the IDB has but 15—and even these staff have no power to influence projects.

MDBs have placed environmental and social guidelines into many of their projects, requiring the borrower countries to inform the local communities of the impending projects; plan and execute viable resettlement projects for affected populations; and mitigate environmental impacts with elaborate zoning rules for development.

Most of these guidelines, however, are impossible for MDBs to enforce unless they stop loan disbursements altogether. MDB critics have even suggested that the banks stop approving loans to a particular borrower until the guidelines for all past loans are followed and small-scale community-based development projects are considered as alternatives to the monstrous schemes.

MDBs have yet to take any of this seriously. First, MDBs will not strain relationships with their borrowers, especially the big ones like Brazil, merely because of environmental or social issues. Nor will MDBs stop the constant flow of capital to which borrower countries have become addicted in order to pay back the MDBs and the commercial banks. If anything, MDBs have been rewarded for environmental destruction by their own donors, who consistently vote to increase MDBs' capital gains.

Five years ago, the explosion of international concern over global warming brought the deforestation issue to the front doorsteps of the industrialized countries as few Third World issues ever have. Activists in the North seized the opportunity to publicize that tropical devastation stems, in large part, from the North's demands for Third World resources, and to enourage concerned citizens to take direct action in defense of the forest and the people who live in them.

One such strategy has focused on persuading donor governments to pressure the multilateral development banks (MDBs) to stop financing ecologically destructive projects, especially those in the rainforests. Started seven years ago in the U.S. by the Washington-based lobbyist groups such as the Environmental Defense Fund, Environmental Policy Institute, and the Sierra Club, the campaign soon grew to include other North American groups, the Rainforest Action Network, the International Rivers Network, Probe International, and scores of other organizations in donor countries in Europe and Japan. The basis of the campaign today is to democratize economic development of the Third World.

"MDBs have a responsibility to fund sound economic development." says Pat Adams, executive director of Probe International. "If they are not doing it, then the taxpayers who bankroll the MDBs should start cutting back on appropriations to these banks. And the people in the South, who ultimately have to live with the environmental and financial consequences of MDB projects, should have the right to shut the projects down and choose their own form of development."

Another front for international campaign work has been to restructure the Tropical Forest Action Plan, an $8 billion ▶

The Rainforest Action Network and the International Rivers Network staged a blockade of the World Bank's 1989 annual meeting in protest of the bank's destructive projects in the rainforest. Photo: J. Majot

The World Rainforest Movement delivers 3 million signatures on a petition calling on the UN to adopt a stronger position on halting tropical deforestation and to stop supporting environmentally destructive programs such as TFAP. Photos: S.Oliveira

program of the World Bank, United Nations, and World Resources Institute (WRI) to manage and develop the world's tropical rainforests, including the Amazon. Environmental activist groups such as the World Rainforest Movement—an international coalition of grassroots support organizations—criticized TFAP so vigorously for excluding local communities in the plan's decision-making process that WRI is now forced to re-evaluate the entire operation.

In the midst of trying to reform the development institutions, northern organizations have not forgotten one of their most powerful weapons in the battle to save the Amazon: the "green shopper." Friends of the Earth, the Rainforest Action Network, and JATAN have launched campaigns to stop the importation and purchasing of tropical hardwoods in the U.S., Europe, and Japan the world's biggest markets—a tactic now endorsed even by mainstream figures such as Prince Charles of Great Britain.

Right now consumer campaigns are our only option to slowing destructive logging, says Pam Wellner, tropical hardwood campaigner of the Rainforest Action Network: "On a large scale there is no sustainable extraction of hardwoods anywhere in the world. Even by conservative standards there is less than 1 percent of any tropical forestry that is considered sustainable. Yet, logging continues and so does the irreparable damage to the forest. The only way to assure that a forest will

Tropical Forest Action Plan— A Blueprint for Disaster

In 1985, First World governments and development banks began touting the multibillion dollar Tropical Forest Action Plan–the blueprint for managing the world's tropical forests–as the solution to the problem of rainforest destruction. Environmentalists, however, have revealed that TFAP will not save the rainforests but lead to their more systematic destruction.

The World Bank, the Food and Agriculture Organization (FAO), the United Nations Development Programme and the World Resources Institute developed TFAP and the forestry department of FAO administers it. TFAP aims to boost investments in tropical forestry as a way of ensuring "the economic and social well-being of rural people in developing countries." It will spend $8 billion to develop forest-based industries, fuel wood production, and conservation and land use management plans.

Some 73 countries containing 85 percent of the earth's tropical forests are lining up for TFAP. Aid agencies are now pouring much of their development assistance funds for Third World forestry activities into TFAP; the United Kingdom, for one, has eagerly pledged almost $100 million.

According to the World Rainforest Movement—an international coalition of nongovernmental organizations—funding for TFAP must be halted until it is completely restructured to address the main causes of deforestation: the reasons for colonization, corruption of forestry officials, destructive development projects and policies, the lack of native rights, and no community involvement in development. Moreover, TFAP intends to intensify logging when no viable plan exists for ensuring sustained-yield practices.

About 20 nations have completed national forest action plans for implementing TFAP, which are now being reviewed by funding agencies. The World Rainforest Movement has conducted in-depth studies of nine national forest action plans, three of which are Amazon nations.

• PERU: The forest action plan adheres to Peru's five-year plan which upholds policies leading to forest destruction such as large-scale development projects and land titling based on land clearing. It promotes deforestation by proposing to increase activities such as agricultural production in rainforest areas (ignoring the problems of soil infertility); road construction; and logging (boosting volumes between 390 and 590 percent).

Omitted from the plan are its impacts on indigenous peoples, and the problems with the drug traffickers and insurgents, the increasing number of Brazilian colonists, the roads linking Brazil and Peru, and conflicts of interest between officials, corporations, and Indians over conservation strategies for national parks.

• COLOMBIA: The forest action plan resembles a wish list of expensive projects, with no details of how they will be executed given the fact that the more accessible forests have already been exhausted by loggers while forest management institutions are in shambles.

The plan does not mention that 19 million hectares of the Colombian Amazon have recently been titled to the Indians. Neither are Indians mentioned in reference to national park expansion and consolidation. The plan calls for 370,000 plantations by the end of this decade, with no discussion of the impacts on local peoples. Logging industries will be made more efficient by supplying them with credit for modern equipment.

• GUYANA: Deforestation is minimal now in Guyana, but the plan calls for a substantial increase in logging to boost state revenues. It claims that the Amerindians are potentially among the greatest beneficiaries of five of the 37 projects. Three of these projects appear possibly destructive to Indian lands, while it is too soon to tell how the other two will be developed. About 62 percent of the $90 million budget will go to logging and plantations through quickly dispersed loans, leaving little time to hire and train staff.

The plan ignores the negative impacts of placer mining on river ecology and the road project linking Brazil and central Guyana, and omits any role whatsoever for community groups. It is not clear who will control reforestation and watershed management projects or how the people dependent on fuel wood will have access to plantation-generated wood. Conservation plans absorb only 7 percent of the plan's budget, and do not refer to land use and land tenure.

exist to implement ecologically sustainable forestry is if we quickly reduce our demand for tropical timber."

In the same vein, support groups around the world are now working to promote the sale of appropriately produced goods from the rainforest. What products to extract, how to market increasing quantities of forest products and still keep the forest intact, and who and where to sell to are decisions that are being made by rainforest communities in consultation with international support groups like Cultural Survival. A small group of anthropologists based in Cambridge, Massachusetts, CS has unexpectedly discovered its own entrepreneurial talents and is putting them to use by developing markets for rainforest products in the U.S. and Europe.

The world's large wildlife organizations also have joined in the chorus denouncing the root causes of deforestation. But in the past year, their conservation strategies for the rainforests have increasingly come under fire by the Amazonian Indians for being "more concerned about the animals than the people who live there." Groups that pump millions of dollars into debt-swap programs to protect "mega-diversity" areas with more park guards and zones of development usually do so without involving the local communities or recognizing their land rights.

"The Indians are asking nongovernmental groups to push the governments of the Amazon to change laws regarding lands," says Maria Ortiz of the Washington-based Conservation International. "I strongly believe that it is not the role of foreign groups to tell the government what to do. It could create a negative response."

What about Latin American-based wildlife groups with whom CI and other U.S. groups work closely? "They still ▶

need to go further," admits Ortiz. "They've started working with journalists and peasants, but still represent the elite and have not gotten to the roots of the deforestation problem."

As a compromise, wildlife groups joined an ongoing committee of Indians and environmental activists that would examine issues of land rights and providing greater financial and technical support for their grass-roots organizing efforts.

Such pledges have been met with a healthy dose of skepticism by Latin American environmental activists such as Gustavo González, a resource management expert with the Ecuadorian environmental coalition Por La Vida. "In Latin America, conservationists' impacts are very limited," says González, "because they never challenge the system, no matter how dangerous or environmentally destructive its policies."

Por La Vida has recently brought together seven Ecuadorian grass-roots environmental groups to work directly with Indians, peasants, and urban dwellers to challenge the legality of Ecuador's destructive developement projects. The coalition has started holding unprecedented public rallies in Ecuador's major cities to gather support for the Amazonian Indians' right to protect their lands from oil and gas development. "The population of Ecuador is 50 percent white and *mestizo*," says González. "If these people are not brought into the process, the needs of the forest peoples and the future of the Amazon will not be considered by the government."

Mobilizing all sectors of society to support the people of the rainforest is one of the most important goals of rainforest organizations, says Randy Hayes, director of the Rainforest Action Network. "It's not enough to keep preaching to the convinced about saving the Amazon. We need to work closer with other movements in the North and South.

"But the bottom line is how effectively we take direction from the forest communities and respond to their requests for help," adds Hayes. "They've been trying to tell us for years. It's just taken many of us a while to hear it." ◆

A demonstration at the Brazilian consulate in San Francisco, California, demanding that the murderers of Chico Mendes be brought to justice.
Photo: ©Aguirre/Switkes

SOUTH AMERICA

▶ **Coordinadora de las Organizaciones Indígenas de la Cuenca Amazónica (COICA)**

(Coordinating Body for the Indigenous Peoples' Organizations of the Amazon Basin)

Objectives: In the last two decades, indigenous leaders of Amazonia have seen an increasing number of governments, banks, and aid agencies make policy decisions about the development of the Amazon. Many of these policies, however, have sent Amazonia barreling toward total devastation. Responding to this crisis, Indian leaders have challenged these institutions on the regional as well as on the international level.

In 1984, COICA was founded to address these issues by the national Indian organizations of five nations. Six years later, COICA has become the international representative of almost all indigenous peoples living in the South American rainforests, and is regarded as the Indian voice of Amazonia today.

"Our people have always been organized, but we have not been listened to," says COICA President Evaristo Nugkuag. "Outsiders have either ignored us or have spoken on our behalf. Now we have a formidable international organization and a common voice. We will argue our own case."

COICA's founding members are the Inter-Ethnic Association for Development of the Peruvian Jungle (AIDESEP), the Central Organization of Ecuadorian Amazonia (CONFENIAE), the Union of Indigenous Nations (UNI), the National Organization of Indigenous Peoples of Colombia (ONIC), and the Central Organizations of Indigenous Peoples of Eastern Bolivia (CIDOB).

COICA's organizational objectives are to defend the rights, territories, and self-determination of indigenous peoples; to represent the national member organizations at international forums; to strengthen the unity of all indigenous Amazon peoples; to promote indigenous cultural values through autonomous development in each country; and to incorporate any interested indigenous people in the process of organizing themselves.

COICA urges conservationists to understand that the best defense of the Amazon is the defense of territories recognized as homelands by indigenous peoples and to respect the leadership of those who have always lived in harmony with the rainforest.

Programs: Internationally COICA is focusing on influencing several powerful decision-making bodies: the United Nations, the multilateral development banks (World Bank and the InterAmerican Development Bank); the Amazon Pact Special Committee on Indigenous Peoples; the International Labor Organization; and major environmental organizations.

In a recent trip to the U.S. in October 1989, COICA delegates started a dialogue with environmental groups about an Indian/environmentalist coalition to save their Amazon homelands. They urged environmental groups to re-examine the idea of the debt-for-nature swap, which is often carried out without involving the Indian organizations.

As an alternative, COICA proposed a "debt-for-Indian stewardship" swap, in which debt would be traded for demarcation and protection of traditional territories. The Indians would play the leading role in the planning and execution of the stewardship swaps and take responsibility for protecting and sustainably developing the new reserves.

Special Needs: Financial support is urgently needed for operations and for building a communications system linking COICA headquarters with its members throughout South America and with support groups in the United States (this includes computers, fax machines, and radios). U.S. contributions to COICA can be made directly or through Oxfam America (page 71).

Resources: Current information on issues related to the Amazon's development and its peoples; and position papers (in English and Spanish).

Contact: Evaristo Nugkuag
Presidente
COICA
Jiron Almagro 614
Lima 11
Perú
Phone: (14) 631-983
Fax: (14) 423-572

In the U.S.:
Contact: Jane Wholey
Esopus Creek Communications
1011 Orleans St.
LA 70116
U.S.A.
Phone: (504) 522-7185
Fax: (504) 522-7185 (call first)

COICA delegation accepting the 1986 Right Livelihood Award (the alternative Nobel Prize) from Jackob Von Uexkull, founder and chairman of the Right Livelihood Foundation. COICA representatives from left to right: José Narciso Jamoijoy (ONIC/Colombia), José Urañavi (CIDOB/Bolivia), and COICA's president, Evaristo Nugkuag (AIDESEP/Peru). Photo: Per Frisk

SOUTH AMERICA
BOLIVIA

Indigenous Organizations

Central de Pueblos Indígenas del Oriente Boliviano (CIDOB)

(Central Organization of Indigenous Peoples of Eastern Bolivia)

Objectives: CIDOB was founded in 1982, uniting approximately 300,000 Indians throughout the Amazon to defend land rights and rationally manage natural resources. "We defend our right to economic and social development; to respect our native language and our culture; and to education consistent with our way of life and the professional training of Indian people," says CIDOB. " We insist upon the right to live with human dignity, and we aim to participate as nations that are part of the Bolivian state, in the definition and implementation of Bolivian economic development policies." CIDOB membership is comprised of 36 ethnic groups from areas including Lomerío, Concepción, Guaraníoes, Ava-Guaraní, Cabildo Mojeño, Santa Rosa del Sara, Ayoreos, and Guarayos.

Programs: CIDOB has established development projects, including two community lumber mills, a small-scale cattle company, training programs in clothing manufacture, communal gardens with womens' organizations and consumers' cooperatives. CIDOB has granted scholarships to Indian university students and conducts a bilingual/bicultural education program with the Guaraní. CIDOB's land protection program is producing a survey of Indian lands that will be used to lobby for changes in government development policies for the Amazon.

Special Needs: Financial assistance for CIDOB's land protection, communications, bilingual education, and womens' programs.

Resources: A monthly bulletin *Oyendo Indígena (Listening to Indigenous Voices)*.

Contact: Miguel García
Presidente
CIDOB
Villa 1ro. de Mayo
Casa Campesina
Casilla 4213
Santa Cruz
Bolivia
Phone: (33) 46-714

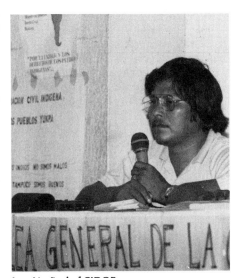

Jose Urañavi of CIDOB.
Photo: ©Aguirre/Switkes

Central de Cabildos Indígenas Moxeños (CCIM)

(Center of Moxeño Indian Councils)

Objectives: CCIM is an organization of the Moxeño, an Indian tribe in the Beni region, devoted to protecting Moxeño rights to their lands and to participating in the planning of regional development schemes by the government.

Programs: CCIM has been an outspoken critic of the Beni Reserve debt-for-nature swap and its negative impacts on indigenous land use and the environment.

Contact: CCIM
Casilla 58
Trinidad, Beni
Bolivia

Comité Ad Hoc de Zafreros Empatronados de la Castaña

(Ad Hoc Committee of Nut Gatherers)

Objectives: The committee formed in 1989 to organize Amazonian nut gatherers to seek better prices for their products and working conditions.

Programs: They plan to lobby for improvements in labor laws, to form a national association of nut gatherers to improve domestic and international marketing techniques, and work more closely with rubber tappers associations.

Contact: Lorenzo González
Presidente
Comité Ad Hoc de Zafreros Empatronados de la Castaña
Calle Progreso 355
Riberalta
Bolivia
Phone: 82-257

Farmers

Central Obrera Departamental (COD)

(Central Workers' Organization)

Objectives: COD is a 40-year-old organization of workers, farmers, and students fighting for social, political, and economic justice, and environmental protection.

Programs: With the University of Santa Cruz, COD has set up training schools for organizing unions and social movements. It also conducts research, lectures, conferences, and seminars on workers rights, economic development, and environmental problems.

Special Needs: Financial support for publishing; broader contacts with interested organizations.

Resources: COD produces a daily television program in Santa Cruz called *Tribuna Popular (Popular Tribune)*, the newspaper *Tribuna Popular*, and the journal *Cuadernos Populares (Popular Notebooks)*.

Contact: Jesús Yavarí
Secretario Ejecutivo
COD
Calle Republiquetas no. 54
Casilla no. 1977
Santa Cruz
Bolivia
Phone: (33) 43-758

▶ **Confederación Sindical Unica de Trabajadores Campesinos de Bolivia (CSUTCB)**
(Confederation of Union Peasant Workers of Bolivia)

Objectives: CSUTCB is a national confederation of all peasant labor unions, which coordinates and supports regional union activities, such as the coca growers' union campaign to uphold Bolivian farmers' right to grow coca and to condemn U.S. eradication efforts.

Programs: CSUTCB is proposing a new plan for agrarian reform, which would help cut down on the flow of landless to the rainforests. CSUTCB is now working more closely with the Indian confederation (CIDOB) on Amazonian land use issues.

Contact: Juan de la Cruz Villca
Secretario Ejecutivo
CSUTCB
c/o CEDOIN
Casilla 11593
La Paz
Bolivia
Phone: (2) 364-975

▶ **Cooperativa Agrícola Integral "Campesino"**
(Cooperative of Agricultural Peasants)

Objectives: Founded in 1980, the Cooperative offers rubber tappers, nut gatherers, and peasant farmers marketing alternatives to the inequitable systems dominated by land owners and traders.

Contact: Víctor Hugo Vaca
Presidente del Directorio
Cooperativa "Campesino" Ltda.
Riberalta, Beni
Bolivia

▶ **Support Groups**

▶ **Asociación Ecológica del Oriente (ASEO)**
(Ecological Association of Eastern Bolivia)

Objectives: ASEO is a conservation organization devoted to protecting the Amazon by promoting ecologically sound natural resource management. ASEO educates the public and scientific communities on the need to protect the rainforest from inappropriate agriculture, logging, and pesticide use.

Programs: ASEO monitors government enforcement of environmental laws; holds a public conference every month on ecological themes; and has formed study commissions to recommend management policies for the following six categories: flora and fauna, land-use patterns, animal husbandry and farming, forestry, and pollution. ASEO is planning to publish an environmental bulletin for peasants.

Special Needs: Funding for institutional support.

Resources: The bimonthly bulletin *Enfoque Ecológico (Ecological Focus)* and *Proceedings of the Ecological Symposium on the Bolivian Tropics.*

Contact: Walter Henry
Presidente
ASEO
Calle Ballivian 157,
Casilla 434
Santa Cruz
Bolivia
Phone: (33) 39-252 or 39-607

▶ **Ayuda para el Campesino Indígena del Oriente Boliviano (APCOB)**
(Aid for the Indigenous People of Eastern Bolivia)

Objectives: Formed by social scientists in 1979, APCOB is the principal support organization for CIDOB, the confederation of Bolivia's Amazonian Indians, on development, education, and communications projects. APCOB recognizes the ancestral rights of Indian peoples to their land, organizations, languages, and cultures. And it proposes "a re-examination of policies of assimilation of Indian peoples to the national society as peasants."

Programs: APCOB lends economic support and technical advice on management of natural resources to indigenous communities. It has also helped create agricultural experimental centers and a horticultural program for several women's organizations.

Special Needs: Financial support for a census of Indian peoples and for a program on managing tropical forests in the Bolivian Amazon.

Resources: The monthly *Boletín del Campesino Indígena (Indigenous Peoples' Bulletin)*; a monthly bulletin; video cassettes *Los Ayoreode–A Latin American History* (about a group of nomadic Indians first contacted in 1948) and *Zapoco, un Proyecto Integral (the Zapoco Project).*

Contact: Dr. Juergen Riester
Coordinador
APCOB
Villa 1 de Mayo s/n
Casilla 4213
Santa Cruz de la Sierra
Bolivia
Phone: (33) 46-714 or 34-986
Fax: (33) 36-404
Telex: 4220 Cab. Pub.

APCOB helped the Ayoréode Indians build a new settlement in the rainforest after their homelands were destroyed by deforestation. Photo: Ted MacDonald/Cultural Survival.

▶ **Centro de Investigación y Documentación para el Desarrollo del Beni (CIDDEBENI)**

(Research and Documentation Center for the Development of the Beni Region)

Objectives: CIDDEBENI is an information and research center on the Beni region's natural resources and Indian populations. It also plans alternative development strategies and builds public support for Indian rights.

Programs: CIDDEBENI is now assessing the social and economic condition of the Moxeño Indians, land-use policies affecting the Chimanes forests, and the lumber industry in the Beni region. CIDDEBENI gives lectures and workshops on Indian rights, natural resource use, conservation and development to the public, scientists, and policy makers.

Special Needs: Funding for operations; international support for Indian land rights; and technical assistance for natural-resource management, especially agroforestry.

Resources: The book *Nuestro Bosque de Mañana (Our Future Forest)*; a series of reports *Los Recursos Forestales en Bolivia (Forest Resources in Bolivia)*, and *Organización y Demandas Indígenas en el Area del Bosque de Chimanes (Indigenous Organizations and Demands in the Area of the Chimanes Forest)*; a slide show on forest preservation, indigenous peoples, and logging.

Contact: Carlos Navia Ribera
Director
CIDDEBENI
Calle Fabián Monasterio
Claure No. 25
Casilla 159
Trinidad, Beni
Bolivia
Phone: (46) 22-824

▶ **Centro de Investigación y Promoción del Campesino (CIPCA)**

(Center of Research and Promotion of Peasants)

Objectives: CIPCA formed in 1971 to support Bolivian peasant movements for self-reliance and economic development.

Programs: CIPCA helps peasant farmers in the Santa Cruz area develop and market locally produced goods. It also conducts educational and radio training programs.

Photo: ©Aguirre/Switkes

Special Needs: Funding for research and qualified personnel.

Resources: Project reports

Contact: Marcos Recolons
Director Nacional
CIPCA
Casilla 5854
La Paz
Bolivia
Phone: (2) 360-5421, 363-440, 374-701
Fax: (2) 391-364

▶ **Coordinadora de Solidaridad con los Pueblos Indígenas del Oriente Boliviano — Asamblea Permanente de Derechos Humanos**

(Coordinating Committee in Solidarity with the Indigenous Peoples of Eastern Bolivia—Permanent Assembly on Human Rights)

Programs: The Committee gives legal assistance to Indian groups on conducting human rights campaigns.

Contact: Sonia Brito
Coordinadora de Solidaridad con los Pueblos Indígenas del Oriente Boliviano
Casilla 1607
La Paz
Bolivia
Phone: (2) 350-486

▶ **Fundación Amigos de la Naturaleza (FAN)**

(Friends of Nature Foundation)

Objectives: FAN is a new organization established to protect biodiversity in Bolivia. FAN was originally a small grassroots organization but has grown to take on large-scale conservation projects in partnership with the Nature Conservancy. FAN's goal is to have at least 10 percent of Bolivia's most significant wildlife habitats designated as protected areas.

Programs: FAN's intial efforts will focus on protecting Amboro and Noel Kempff National Parks by the training and equipping of park guards.

Resources: FAN newsletter

Contact: FAN
Av. Irala 421
P.O. Box 300
Santa Cruz
Bolivia
Phone: (33) 44-820 or 50-544
Fax: (33) 35-601
Telex: 4210 TELECT BV

▶ **Instituto de Documentación y Apoyo Campesino (IDAC)**

(Institute of Documentation and Support for Peasant/Indian Peoples)

Objectives: IDAC lends legal and administrative support to Indigenous peoples' struggles for land and human rights.

Programs: IDAC investigates the terms of land deals offered to Indians by speculators and corporations. IDAC also helps Indians and peasants of the Camire, Santa Cruz area resettle in more productive agricultural areas.

Special Needs: Funding for operations and an office building.

Resources: Project reports, photos.

Contact: Franz Michel
Director
IDAC
Cll. Rene Moreno 58
Casilla 43
Camiri, Prov. Cordillera
Santa Cruz de la Sierra
Bolivia
Phone: (33) 2453

Liga de Defensa del Medio-Ambiente (LIDEMA)
(Environmental Defense League)

Objectives: LIDEMA is a national network of eleven environmental, scientific, and public policy organizations that formed in 1985 to coordinate education and research projects on Bolivian environmental problems and to fight for the creation of wildlife reserves.

Programs: LIDEMA was instrumental in the creation of the Beni Biosphere Reserve and the creation of a national park system. LIDEMA has recently drawn up the management plan for the Beni Reserve's biological station and the integrated development of the Chimanes Forest. LIDEMA also gives workshops to policy makers, journalists, and scientists on its programs.

Special Needs: Funding for conservation programs, management of protected areas, and for the National Museum of Natural History. LIDEMA invites self-funded researchers to join its programs.

Resources: The quarterly bulletin *Lidema*; the reports *Perfil Ambiental de Bolivia (Environmental Profile of Bolivia)*, *Manual de Organización y Participación Comunitaria (Manual of the Organizational Process and Community Participation)*, and *Diagnóstico de Diversidad Biológica de Bolivia (Diagnosis of Biological Diversity of Bolivia)*; a library; videos.

Contact: Carlos Landívar
Director Ejecutivo
LIDEMA
Av. 20 de Octubre No. 1763
Casilla 11237
La Paz
Bolivia
Phone: (2) 356-249 or 353-352

Resource Groups

Centro de Datos para la Conservación (CDC)
(Conservation Data Center)

CDC is one of ten such centers created in Latin America and the Caribbean with the support of the Nature Conservancy (U.S.) to provide data on biodiversity, land use, and other ecological subjects for scientists and the government agencies.

Contact: Dra. María R. de Marconi
Directora Ejecutiva
CDC
Ave. 6 de Agosto No. 2376
Casilla 11250
La Paz
Bolivia
Phone: (2) 352-071
Fax: (2) 352-071 (call first)

Centro de Documentación e Información (CEDOIN)
(Center of Documentation and Information)

CEDOIN is an alternative information center on issues concerning the economy, politics, the military, drug trafficking, and the environment.

Special Needs: Funding for operations and for publishing *Informe Rural* in eastern Bolivia.

Resources: The bulletin *Informe "R"* (five times/yr); the bimonthly *Bolivia Bulletin* (in English); the monthly *Informe Rural (Rural Information)*; *Informes Especiales (Special Bulletin)* 5 times/yr; books and newspaper clippings.

Contact: Sara Monroy
Directora Ejecutiva
CEDOIN
Av. Montes # 710 3er piso
La Paz
Bolivia
Phone: (2) 372-940

Centro de Estudios para el Desarrollo Laboral y Agrario (CEDLA)
(Center for the Study of Labor and Agrarian Development)

CEDLA helps urban and rural communities plan and execute small-scale development projects. In the Amazon, CEDLA has helped organize the Campesino Cooperative, the Federation of Peasants of Beni and Pando, and the union of rubber tappers and nut collectors.

Resources: An information center; the reports *Beni y Pando Latifundio y Minifundio (Beni and Pando, Latifundia, and Minifundia)* and *Amazonía Boliviana y Campesinado (Bolivian Amazon and Peasants)*.

Contact: Dr. Antonio Péres Velasco
Director
CEDLA
Casilla 8630
Pasaje Aspiazu (743)
No. 2032
La Paz
Bolivia
Phone: (2) 340-746 or 360-223

Misión Británica de Agricultura Tropical
(British Tropical Agricultural Mission)

The Mission works with frontier settlers to develop techniques and technologies for sustainable alternatives to migratory slash-and-burn agriculture. It also works on new technologies for limiting ecological damage of denuded areas.

Special Needs: Information on appropriate agricultural technologies.

Resources: Project studies.

Contact: John Wilkins
Project Head
MBAT
Casilla 359
Santa Cruz
Bolivia
Phone: (33) 46-556
Telex: 4222 BTAM BB

Tatu Gigante by LIDEMA

SOUTH AMERICA
BRAZIL

▶ **Aliança dos Povos da Floresta**
(Alliance of the Peoples of the Rainforest)

Objectives: For many decades, Indians, rubber tappers, and farmers have been at odds with each other over land use in the Amazon. But in the past five years these same groups have found that they all share a common enemy: large-scale environmentally and socially devastating development projects. If the present rate of construction of hydro dams, mining, and ranching projects continues, none of these groups will have a home left in the rainforest.

In March 1989, the national meeting of rubber tappers in Rio Branco, Acre, called on Amazonian Indians and other forest communities to put aside their hostilities and suspicions of each other and declare an alliance to protect the rainforest.

Declaration of the Peoples of the Forest:

"The traditional populations that today mark in the sky of Amazonia the Rainbow of the Alliance of the Peoples of the Forest proclaim their intention to preserve and to remain in their region. They understand that the development of the potential of these populations and of the regions in which they live make up the economic future of their communities, and must be maintained by the entire Brazilian nation as a part of their identity and pride.

"This Alliance of the Peoples of the Forest—bringing together Indians, rubber tappers, and riverine populations, starting here in this region of Acre—extends its arms to embrace all possible strength for the protection and preservation of this immense but fragile system of life that includes our forests, lakes, rivers, and streams—the source of our riches and the basis of our cultures and traditions."

Programs: The Alliance has created the *Embaixada dos Povos da Floresta* (Embassy of the Peoples of the Forest) with a library, a video, photographic and music collection, and art on the cultures and popular movements in the Amazon. The embassy is located in Caxingui, in São Paulo, on land given by the city government to UNI in 1989.

Contact: Embaixada dos Povos da Floresta
Praça Enio Barbato s/n
05.517 São Paulo, SP
Brazil
Phone: (11) 211-9996
Fax: (11) 211-9996 (evenings)

▶▶▶ **Indigenous Organizations**

▶ **Associação de Mulheres do Alto Rio Negro (Numiã-Kuruã)**
(Association of Women of the Upper Rio Negro)

Objectives: Numiã-Kuruã formed in 1984 to assist Indian women of the Rio Negro who live in Manaus. The military and the church brought many of these women to the city to work as domestics, but because of illness or lack of jobs, many have been left destitute and homeless.

Programs: Numiã-Kuruã provides Indian women living in Manaus with medical care, legal assistance on labor issues, and training in health, education, and crafts marketing. Numiã-Kuruã also works with other indigenous organizations of the Rio Negro for the demarcation of their lands.

Special Needs: Financial support for national meetings and to purchase raw materials for handicraft production, such as tucum leaves collected by the Indian women from the forests of the upper Rio Negro and thread made by the Tikuna Indians.

Contact: Deolinda Freitas Prado
Coordenadora
Numiã-Kuruã
Caixa Postal 817
69.000 Manaus, AM
Brazil
Phone: (92) 244-2480

▶ **Associação Rural das Comunidades Indígenas do Baixo Rio Negro (ARCIBRN)**
(Rural Association of Indigenous Communities of the Lower Rio Negro)

Programs: ARCIBRN trains nontraditional Indian communities of the lower Rio Negro in agriculture, animal husbandry, and fisheries.

Contact: Braz de Oliveira França
Presidente
ARCIBRN
Av. Castelo Branco no. 232
69.750 São Gabriel da Cachoeira, AM
Phone: Brazil

The first meeting of the Peoples of the Forest in 1989 with poster of Chico Mendes. From left: Raimundo Barros, Chico Ginú, and a representative of the Union of Indigenous Nations.
Photo: Kit Miller

Casa do Estudante Autóctone do Rio Negro (CEARN)

(Native Students' House of the Rio Negro)

Objectives: Founded in 1985, CEARN is comprised of 60 Indian students from various tribes, including the Tukano, Tariano, Yanomami, Arapasso, Macu, and Pira-Tapuia. It works to help maintain their values and traditions, and to participate in the fight for land rights and cultural survival.

Programs: CEARN hosts Indian students from Amazonas state who are studying in technical schools or at the University of Amazonas and helps them with registration and other bureaucratic procedures.

Special Needs: Funding for outreach to Indian groups and the collection of interviews on Indian traditions.

Resources: An information center on Indian culture, land rights, and history.

Contact: Ismael Pedrosa Moreira
CEARN
Caixa Postal 169
69.011 Manaus, AM
Brazil

Conselho Indígena de Roraima (CIR)

(Indigenous Council of Roraima)

Objectives: CIR was formed in 1987 at an assembly of Indian leaders from the Macuxí, Wapixana, Taurepang, and Ingariko groups to help implement community development projects; to strengthen Indian identity and cultural values, language, and customs in Indian schools; to pressure the government to demarcate Indian lands; and to build public support for Indian rights.

Programs: Among CIR's many community projects are cattle breeding, supply houses and gardens, and the manufacture of clothing.

Special Needs: Funding and wider distribution of their publications.

Resources: The monthly newsletter *Vale a Pena Ler (Worth Reading)*.

Contact: Terêncio Luís Silva
Coordenador
CIR
Rua Sebastião Diniz
Nº 1672 W
69.300 Boa Vista, RR
Brazil
Phone: (95) 224-5761

Suruí warriors.
Photo: ©Aguirre/Switkes

Federação das Organizações Indígenas do Rio Negro (FOIRN)

(Federation of Indigenous Organizations of the Rio Negro)

Objectives: FOIRN was formed by Indian communities of the Rio Negro area to work for the demarcation of Indian reserves in order to protect their lands against Calha Norte, a massive military project that would open the entire region to military and industrial development.

Programs: FOIRN organizes Indian communities, principally those affected by the Calha Norte, to legally challenge the military's expropriation of their land and to demand demarcation of their territories.

Contact: Orlando Melgueiros
FOIRN
Caixa Postal 31
69.750 São Gabriel da Cachoeira, AM
Brazil

Organização do Conselho Indígena Munduruku (OCIM)

(Organization of the Munduruku Indigenous Council)

Programs: OCIM is a community organization of the upper Madeira River region working in education.

Contact: Frei José Marcaloni
OCIM
Aldeia Indígena Kwata
Paroquia N. Senhora de Nazaré
69.230 Nova Olinda do Norte, AM
Brazil

Organização Metareila do Povo Indígena Suruí

(Metareila Organization of the Suruí Indian People)

Objectives: First contacted by outsiders in 1969, the Suruí Indians of Rondônia formed Metareila to cope with the rapid loss of their lands and health to colonization, particularly from the Polonoroeste resettlement scheme. The Suruí must now combat an alarming number of cases of tuberculosis, influenza, smallpox, and meningitis introduced by settlers. They are also trying to avoid the rapid disintegration of their traditional culture.

"We must avoid further suffering of our people," says José Itabira Suruí, Metareila's president. "The lack of help from the government has forced us to do this work ourselves."

Programs: Metareila's health project is building a clinic and hiring medical staff. Its education project plans to build schools and hire teachers, and its development program is creating marketing strategies for rubber, Brazil nuts, fruits, and handicrafts. The Suruí are also working with other Indian groups in Rondônia to help each other defend their territories.

Special Needs: Immediate funding for its projects and for vehicles to transport patients with infectious diseases out of the communities and into the clinics; to transport agricultural produce for marketing; and to protect the Suruí reserve against illegal settlers, hunting, fishing, and logging. A Hand Cam M8 video camera would be especially useful to document the group's activities for wider distribution.

Contact: José Itabira Suruí
Presidente or
João Hilário
Administrador
Metareila
Rua Capitão Rui Teixeira
no. 1666
Caixa Postal 330
78.935 Riozinho-Cacoal
RO
Brazil
Phone: (69) 441-3148

União Das Nações Indígenas (UNI)
(Union of Indigenous Nations)

Objectives: UNI is the national organization of indigenous peoples of Brazil, founded in 1980 to guarantee Indian peoples' rights to their traditional territories, to protect these lands against destruction, and to maintain their ecological balance by promoting sustainable use of natural resources. In the 1980s, UNI established contacts with most of Brazil's 180 Indian nations, and has played an increasingly vital role in organizing Indian opposition to government policies that will have damaging impacts on people and environment.

Programs: UNI works for Indian rights legislation at the national level. It coordinated the Indian lobby at the Brazilian Congress during the writing of the Indian rights chapter in Brazil's new constitution. Through the media, UNI garners broader public support for Indian issues and cultural preservation. In December 1989, UNI National Coordinator Ailton Krenak and Kayapó leaders focused national media attention on the government's neglect of Yanomami Indians.

Through its weekly radio program, UNI has also opened up communication between native peoples and the general public. The program, which is a forum to discuss all issues pertinent to Indians, is broadcast in São Paulo and distributed on cassette to hundreds of Indian communities, who can record their responses and other messages and send them back to the program for airing.

UNI has established a Center for Indian Research and Training on Resource Management, dedicated to preserving traditional knowledge about the Amazon forest and applying it to sustainable development. The Center intends to incorporate into its development projects Western methods and technology that have formerly been denied Indians. But it will do so selectively as not to compromise traditional systems.

"If we are not very careful about the path we take," says Krenak, "and we become dependent on consumerism, then in the future there won't be any difference between indigenous people and any other people. I feel that what kept our people on their feet until today—what maintained the rainforest in which our people live—is the spirit of nature and the invisible spirit of our people. If we're not alert, this will be broken, and we will be lost."

The Center, located in Goiânia, will include:

▲ An integrated project on forest management and sustainable agriculture to grow native fruits and perennial crops and sell them to domestic and international markets; to replenish endangered species populations; and to raise and market traditional livestock.

▲ An inventory of natural resources in Indian reserves to determine the products available for sustainable use and to identify denuded areas in need of regeneration.

▲ Centers for food processing and distribution.

▲ Pilot development projects established in Indian communities based on the Center's model, but adapted to and controlled by each community. Pilot projects are now underway in the Xavante and Suruí areas.

▲ Training Indians to run the center and its projects. Students from eight Indian tribes who will work at the Center are now enrolled in biology and law programs at the Catholic University of Goiás.

Special Needs: UNI needs financial and technical support for the Center; information on similar sustainable development projects; and compact, broadcast-quality tape recorders and dubbing equipment for the radio network. Donations can be made directly to UNI. For information contact Beto Borges at the Rainforest Action Network (see RAN page 72).

Resources: *Programa do Indio*, a radio program broadcast weekly in São Paulo; *Jornal Indígena*, an occasional newsletter.

Contact: Ailton Krenak
National Coordinator
UNI
Praça Enio Barbato s/n
05.517 São Paulo, SP
Brazil
Phone: (11) 211-9996
Fax: (11) 211-9996 (evenings)

Indigenous delegations at Brazil's Constitutional Assembly, 1988.
Upper left: UNI's president, Ailton Krenak. Above: Kayapó delegates.
Photos: ©Aguirre/Switkes

▶ **Regional UNI Councils**

UNI's national policies and programs are determined by regional councils, which are made up of tribal council representatives. UNI has three regional offices, two of them are in the Amazon.

UNI-Acre: UNI-Acre trains native people to work in the health and education facilities in Acre's Indian communities. It also operates cooperatives that market rubber and traditional crafts.

Contact: Antonio Apurinã
UNI-Acre
Rua Francisco Ferreira da Silva, 68
Baixa da Colina, Cohab do Bosque
69.900 Rio Branco, AC
Brazil
Phone: (68) 224-5973

UNI-Amazonas: This council, representing tribes affected by the military's Calha Norte project, has worked to stop mining companies and gold panners from operating on Indian lands. UNI-Amazonas also has a project to identify marketable forest products.

Special Needs: Funding for all programs.

Contact: Manuel Fernandes
Moura Tukano
UNI-Amazonas
Caixa Postal 3774
69.011 Manaus, AM
Brazil

▶ **União das Nações Indígenas Regional Tefé**

(Regional Union of Indigenous Nations of Tefé)

Programs: The Tefé group is an indigenous organization with no direct links with UNI. Its work is mainly in developing community gardens and programs in education and health.

Contact: Miguel
União das Nações Indígenas Regional Tefé
Praça Santa Tereza, 284
69.749 Tefé, AM
Brazil

▶ **Forest Dwellers**

▶ **Associação Das Comunidades Remanescentes de Quilombos do Município de Oriximiná**

(Association of Communities Remaining from Colonies of Escaped Slaves in the Municipality of Oriximiná)

Objectives: The Association formed in the last few years to fight eviction of its members from their homes to make way for large-scale development projects. Members of the Association are the descendants of African slaves who escaped from the estates near Santarém and Obidos and formed *quilombos*, or colonies, in the highlands above the rapids. By the 20th century, many of the *quilombos'* members had migrated to the lower riverine areas of Oriximiná. Since 1970, many of these families have been expelled from their lands when the Trombetas Biological Reserve and the mining area of Mineração Rio do Norte were created.

Programs: The Association is fighting to stop the construction of the Cachoeira Porteira Dam, which would flood 620 square miles of *quilombos* lands. They have filed suit to stop the dam's construction.

Contact: Associação das Comunidades Remanescentes de Quilombos do Município de Oriximiná
a/c Padre José
Oriximiná, AM
Brazil

▶ **Comissão em Defesa do Rio Uatumã**

(Commission in Defense of the Uatumã River)

Objectives: The Commission was created in April 1989 by the people of the Uatumã River to demand government redress of the devastating environmental and social impacts of the Balbina dam's construction.

Programs: The Commission has organized public demonstrations in Manaus demanding compensation for the loss of native fisheries and fresh water supplies, and for medical treatment of health problems caused by Balbina's pollution.

Special Needs: The commission needs financial assistance and international support.

Contact: Maria Eunice Barbosa da Silva
Comissão em Defesa do Rio Uatumã
a/c Prelazia de Itacoatiara
Itacoatiara, AM
Brazil

Riverine people of the Uatumã assess Balbina dam's pollution downstream.
Photo: ©Zbigniew Bzdak

Tucuruí dam. Photo: ©Aguirre/Switkes

▶ **Comissão Regional de Atingidos Por Barragens (CRAB)**

(Regional Commission of Dam Victims)

Objectives: CRAB was founded in 1979 to defend the land and civil rights of people impacted by the Brazil's Plan 2010, which calls for the construction of more than 130 large hydroelectric dams in Brazil—70 of which are targeted for the Amazon. These dams would flood tens of thousands of square miles of the rainforest and the homes of thousands of Indians, farmers, fishermen, and others who live in the forest.

Programs: CRAB held the First Regional Encounter of Workers Affected by the Xingu Hydroelectric Complex in February 1989. It brought together rural unions and other grassroots groups for the first time to discuss the concerns about the resettlement plans and impacts. The meeting was followed in March 1989, with state-wide meetings in Amazonas and Rondônia.

In April 1989, CRAB held a national meeting in Goiânia that brought representatives of 26 Indian groups, farmers and labor to build solidarity between all dam victims in Brazil. Participants set priorities for the coming years, which included building a strong network between affected communities to share information; lobbying the government; and forming an advisory group. The resolutions of the meeting, contained in the *Letter of Goiânia*, called for the government to:

▲ Re-evaluate its energy policy with the help of affected workers.
▲ Solve the social and environmental problems caused by dams already constructed before starting new projects.
▲ Comply with past agreements negotiated between the victims of dams and the electric companies on resettlement.
▲ End federally subsidized electricity rates for certain industries.
▲ Develop energy alternatives.

Special Needs: Financial assistance and information on energy alternatives and the impacts of high dams.

Resources: A monthly bulletin *A Enchente do Uruguai (The Flooding of the Uruguay River)*; a pamphlet called *Terra Sim, Barragens Não (Land Yes, Dams No)*; documents of the National Meeting of CRAB; a video now in production; photographic archive; educational pamphlets.

Contact: Luiz Alencar Dalla Costa
Secretário Geral
CRAB
Rua Bahia, 78
Caixa Postal 372-99700
Erexim, RS
Brazil
Phone: (54) 321-3304

CRAB is working with these groups:

▶ **Comissão Regional dos Atingidos pelo Complexo Hidrelétrica do Xingu (CRACOHX)**

(Regional Commission of Victims of the Xingu Hydroelectric Complex)

Contact: Benedito do Prado
CRACOHX
Sindicato dos
Trabalhadores Rurais
68.370 Altamira, PA
Brazil

▶ **Comissão dos Atingidos por Tucuruí (CATU)**

(Commission of Victims of Tucuruí)

Contact: Aida
CATU
Comissão Pastoral da Terra
68.460 Tucuruí, PA
Brazil

▶ **Atingidos de Ji-Paraná**

(Victims of Ji-Paraná Dam)

Contact: Atingidos de Ji-Paraná
a/c Sindicato dos
Trabalhadores Rurais
Ji-Paraná, RO
Brazil

▶ **Atingidos de Cachoeira Porteira**

(Victims of Cachoeira Porteira Dam)

Contact: Atingidos de Cachoeira
Porteira
a/c Padre José
Oriximiná, AM
Brazil

▶ **Comitê de Defesa da Ilha de São Luis**

(Committee in Defense of the Island of São Luis)

Objectives: The Committee formed to fight the relocation of 25,000 people living near the Alumar aluminum plant, which is owned by Alcoa and Royal Dutch Shell and part of the Greater Carajás mining project.

Contact: José de Nascimento
Comitê de Defesa
da Ilha de São Luis
Morais Filho
Bico do Couto 56 — Centro
65.010 São Luis, MA
Brazil
Phone: (98) 222-3457

Conselho Nacional Dos Seringueiros (CNS)
(National Council of Rubber Tappers)

Objectives: The CNS was created at the First National Encounter of Rubber Tappers in Brasília in 1985, by 130 participants to improve the rubber tappers' social and economic development.

An integral part of CNS's platform is to save the rainforest from destructive practices (cattle ranching, road building, and giant colonization schemes) so that it can survive and continue to support the people who live in it sustainably.

"We are not anti-development," assured CNS founder Chico Mendes shortly before his death. "You only need to see how poor we are to know that."

Mendes' legacy is a program that has become the world's best-known blueprint for sustainable development in tropical forests. Its centerpiece is the establishment of extractive reserves—an agrarian reform measure in which large rainforest areas become state owned but managed by the rubber tappers and exploited only by forest collectors (rubber tappers, fruit and nut gatherers, Indians, and fishermen). Their activities will allow for regeneration of the forest rather than its destution.

Rubber tapper gathering latex. Photos: © Aguirre/Switkes.

Rubber tapper *empate* tears down a shack built by forest clearing crews.

The CNS is also working to:
▲ Defend the Amazon forest and stop the violence against rubber tappers by developers.
▲ Shift government priorities away from road building, cattle ranching, and other large-scale development schemes to decentralized community-based projects.
▲ Improve marketing capabilities, health, and education within the extractive reserves.
▲ Build an alliance of all peoples of the forest with environmentalists, scientists, and human rights advocates.
▲ Gain international recognition from governments, aid agencies, and banks.

Programs: CNS's work with environmental groups in the U.S. and Europe have helped create international support from the public, the governments, and the banks for extractive reserves. Today there are 19 extractive reserves. To stop more destructive development projects, CNS representative Osmarino Amâncio Rodrigues went to Washington in January 1990 to lobby the World Bank against approving a new loan for the Polonoroeste colonization project in Rondônia. CNS has also established cooperatives and new markets for their products, as well as schools and clinics.

Special Needs: Financial support and technical and scientific assistance. "We need people who can help us evaluate our natural resources, develop marketing strategies and a basic economic infrastructure," says Osmarino Amâncio Rodrigues. "We also need to develop organizational techniques so that our products can move into the market in a serious way and we can keep some of the value of the products we produce."

Contributions to CNS may be made directly through Account No. 41046/2, at the agency of the *Banco do Brasil*, Rio Branco, Acre, with previous notification to the CNS. Contributions can also be made via the Chico Mendes Fund in the U.S. (see page 75).

Resources: *Chico Mendes*, a collection of interviews in Potuguese with the late rubber tapper leader.

Contact: Júlio Barbosa de Aquino
Presidente
CNS
Rua Guanabara Nº 150
Estação Experimental
69.900 Rio Branco, AC
Brazil
Phone: (68) 226-3041

Sindicato Dos Trabalhadores Rurais

(Union of Rural Workers, Xapuri, Acre)

Objectives: There are hundreds of chapters of the Rural Workers Union (STR) throughout Brazil that are fighting for better working conditions for farm laborers, ranch workers, and rubber tappers. Chico Mendes cofounded the Xapuri STR chapter in 1977 mainly to curb the rapid growth of cattle ranches and end the monopoly that *patrões*, or rubber barons, maintained over rubber prices. Mendes organized workers to find ways to market rubber which could be controlled by the rubber tappers themselves. Mendes was president of STR in Xapuri from 1982 until his death in December 1988.

Programs: The Xapuri Rural Workers' Union organizes *empates*—nonviolent demonstrations to protect the rainforest and organize rural workers. In an *empate*, rubber tappers and their families block bulldozing, logging, and burning operations with their bodies, or they occupy government offices until the authorities cancel land clearance permits. The actions have saved thousands of acres of rainforests in the Xapuri area alone.

The Xapuri Union also runs the literacy campaign called *Projeto Seringueiro* (Project Rubber Tapper) that incorporates the social and cultural issues of rubber tapper communities into the reading and writing curricula. The project has been adopted by 19 schools and is also responsible for training rubber tappers as teachers. More than 1,000 adults have learned to read through this program. Six health posts in the rubber tapping areas have also been established.

Special Needs: Financial assistance and allies. "I'd like to invite people and groups from other countries to visit us," Chico Mendes once said, "to have a close-up look at our reality, because only in this way will they have a deeper understanding of our situation. Only then will they have a firmer basis upon which to fight in our defense."

Resources: A unique collection of Mendes' papers, awards, and photos is maintained at his house in Xapuri.

Contact: Jorge Gomes Pinheiro
Presidente
Sindicato dos
Trabalhadores Rurais
Av. Benjamin Constant
s/n
69.900 Xapuri, AC
Brazil

Movimento de Mulheres Do Campo e da Cidade

(Movement of Women from the Countryside and the City)

Origins: Rural women workers formed this union to fight exploitation and discrimination against them in the eastern Amazon. Over the years, the Movement grew to include housewives, Brazil nut processing workers, and women metalworkers.

Special Needs: Financial support for travel and printing costs. It wants to broaden its contacts with other women's groups, unions, and student organizations.

Contact: Iza Cunha
Presidente
Movimento de Mulheres
do Campo e da Cidade
Rua Manoel Barata 47
sala 305
Centro
66.020 Belém, PA
Brazil

Father Ricardo Rezende works in suppport of landless farmers. Photo: ©Aguirre/Switkes

Farmers

Movimento Dos Trabalhadores Rurais Sem Terra (MST)

(Movement of Landless Rural Workers)

Objectives: MST is a national organization formed in 1985 to fight for land-reform policy benefiting an estimated 14 million landless peasants in Brazil. It is an independent grassroots movement run by commissions representing squatters' camps and settlements.

Programs: MST uses various forms of collective action: street demonstrations, regional assemblies, audiences with government officials, hunger strikes, protest camps in cities, and the occupation of government offices and contested properties. Today some 5,000 landless families are occupying estates in Brazil to force redistribution of land. MST has brought about redistribution of 360 estates, totalling 2.5 million acres.

Special needs: Assistance in developing agricultural cooperatives and solving production problems in the settlements.

Resources: The monthly newspaper *Jornal dos Sem Terra* (*Newspaper of the Landless*); brochures in Portuguese and English; teaching pamphlets.

Contact: Movimento do Sem Terra
Rua Ministro Godoy, 1484
05.015 São Paulo, SP
Brazil
Phone: (11) 864-8977
Telex: (11) 82153 MSTB BR

▶ **Support Groups**

▶ **Ação Democrática Feminina Gaúcha— Amigos da Terra (ADFG)**

(Democratic Feminine Action of Rio Grande do Sul— Friends of the Earth)

Objectives: ADFG started as a womens' environmental group, but is now one of the principal environmental groups in Brazil.

Programs: ADFG works with international environmental movement on toxics, sustainable agriculture, rainforest preservation, reform of multilateral development banks, and changing the international economic order.

Resources: Pamphlets on ecology and recycling.

Contact: Magda Renner
Presidente
ADFG
Rua Miguel Tostes, 694
90.420 Porto Alegre, RS
Brazil
Phone: (512) 328-884

▶ **Ação Ecológica Vale do Guaporé (ECOPORÉ)**

(Ecological Action for the Guaporé Valley)

Programs: ECOPORÉ works to protect the forest and Indian reserves in Rondônia and also conducts environmental education programs in Guaporé Valley.

Special needs: ECOPORÉ needs photographic and video equipment.

Resources: Reports on deforestation covering illegal logging and the use of defoliants.

Contact: João Alberto Ribeiro
ECOPORÉ
Av. Rio Branco, 4226
78.965 Rolim de Moura
RO, Brazil
Phone: (69) 442-1262

▶ **Assessoria e Serviços a Projetos em Agricultura Alternativa (AS-PTA)**

(Support and Services to Alternative Agriculture Projects)

Objectives: PTA was founded in 1983 by agronomists to help peasant farmers find alternative strategies for agricultural and livestock development.

Programs: AS-PTA helps peasant farmers make contact with other communities to exchange new techniques and ideas in production and marketing, such as the formation of cooperatives. AS-PTA coordinates activities of communites in the states of Maranhão, Ceará, Paraíba, Pernambuco, Bahia, Minas Gerais, Espírito Santo, Paraná, Santa Catarina, and Rio Grande do Sul.

Special needs: Funding for operations; information from similar organizations on appropriate technologies.

Contact: Jean Marc von der Weid
AS-PTA
Rua Bento Lisboa 58
3º andar—Catete
22.221 Rio de Janeiro, RJ
Brazil
Phone: (21) 285-2998
Telex: (21) 34201 FOAS BR

▶ **Associação Gaúcha para a Proteção do Ambiente Natural (AGAPAN)**

(Gaúcha Association for the Protection of the Natural Environment)

Programs: Founded by environmentalist José Lutzenberger, AGAPAN's programs protect and restore the quality of the natural and human environment, mainly through sustainable agriculture. AGAPAN's focus has been on southern Brazil.

Contact: Celso Marques
Presidente
AGAPAN
Caixa Postal 1996
90.210 Porto Alegre, RS
Brazil
Phone: (512) 287-352

▶ **Associação de Informações da Amazônia**

(Information Association of Amazônia)

Objectives: The Association was formed by Lúcio Flávio Pinto, Brazil's leading journalist on Amazonian issues, to independently investigate and publish information on the impacts of large-scale projects in his native state of Pará and to lend organizing assistance to forest communities on development projects.

Special needs: The Association is trying to obtain a desktop publishing system (for Portuguese); a high-powered personal computer for the data bank; and broader contacts with others working on Amazonian issues.

Resources: *Jornal Pessoal (Personal Newspaper)*, twice a month; and the book:, *Carajás: o Ataque no Coração da Amazônia (Carajás: The Attack on the Heart of Amazonia)*.

Contact: Lúcio Flávio Pinto
Presidente
Associação de Informações da Amazônia
R. Aristides Lobo, 871
66.030 Belém, PA
Brazil
Phone: (91) 224-3728
Telex: 911033

▶ **Associação Mundial de Ecologia (AME)**

(World Association of Ecology)

Objectives: AME has programs to raise the ecological awareness of the public, especially within the schools.

Contact: Waldemar Paioli
Presidente
AME
Rua Senador Feijó
338 - 2º andar
C.P. 108
06.700 Cotia, SP
Brazil
Phone: (11) 493-2636 or 628-952

▶ **Campanha Nacional pela Defesa e Desenvolvimento da Amazônia (CNDDA)**

(National Campaign for the Defense and Development of Amazônia)

Objectives: CNDDA was formed in 1967 by geographer Orlando Valverde to protect the natural resources and inhabitants of Amazonia and to promote conservation and appropriate development.

Programs: In conjunction with 14 other organizations, CNDDA has brought suit to stop charcoal producers of Carajás from using up vast portions of the surrounding rainforest.
CNDDA is sponsoring research and development of activated charcoal filters that will guard against mercury poisoning in the gold-panning process.

Special Needs: Financial support, information exchange, and assistance with publicity on Amazonian issues.

Resources: The journal *Amazônia Brasileira em Foco* (*The Brazilian Amazon in Focus*); the bulletin *Amazônia—Informação e Debate* (*Amazonia—Information and Debate*) three times/yr; audio-visual materials.

Contact: Orlando Valverde
Presidente
CNDDA
Rua Araújo Porto Alegre
71/ 10º andar
20.030 Rio de Janeiro, RJ
Brazil
Phone: (21) 262-5734,
282-1292 ext. 26
275-1057, or 240-8769

▶ Campanha Nacional pela Reforma Agrária (CNRA)

(National Campaign for Agrarian Reform)

Objectives: CNRA was created in 1983 to build public support, especially in the cities, for agrarian reform. CNRA is made up of 56 organizations, including indigenous groups, labor unions, professional and scientific associations, environmental and church groups.

Programs: CNRA is investigating land ownership concentrations in Brazil; campaigning to stop the use of violence against workers; and lobbying the Brazilian Congress for legislation protecting rural workers.

Special Needs: CNRA needs financial support, information exchange on land reform, international support for agrarian reform.

Resources: The bimonthly bulletin *Informa* (*Inform*); videos, posters, and publications on agrarian reform.

Contact: Herbert de Souza
CNRA
Rua Vicente de Souza, 29
22.251 Rio de Janeiro, RJ
Brazil
Phone: (21) 286-0348
Fax: (21) 286-0541
Telex: 021-36466 BASE BR
Econet: IBASE
Alternex: 7241212047902

▶ Centro de Defesa dos Direitos Humanos (CDDH)

(Center in Defense of Human Rights)

Objectives: CDDH is a organization of the Catholic Church that works primarily to protect the human rights of urban social movements.

Programs: CDDH provides legal assistance on human-rights cases; produces educational materials; and has also set up a permanent national forum on human rights.

Special Needs: Assistance with computer training; financial support for law students specializing in human rights work; and closer links with other human rights groups.

Contact: Aparecida Maria
CDDH - CNBB Norte 1
Av. Epaminondas,
722 - Centro
69.010 Manaus, AM
Brazil
Phone: (92) 233-8072

▶ CDDH Regional Office

CDDH-Acre: This office works with grassroots communities, women's groups, rural unions, rubber tappers, leprosy victims, and environmental groups in Acre.

Contact: Carlos Hideli Kawahara
Coordenador
CDDH-Acre
Trav. Cabanelas, 40
69.900 Rio Branco, AC
Brazil
Phone: (68) 224-5772

Migrant settler working the infertile soils of the Amazon.
Photo: ©Aguirre/Switkes

▶ Centro Ecumênico de Documentação e Informação (CEDI)

(Ecumenical Center For Documentation and Information)

Objectives: During Brazil's military dictatorship, a group of social scientists with close ties to the ecumenical community formed CEDI to help build support for democratic movements. CEDI's professional teams conducts research and produces a multitude of publications on social and political issues critical to native peoples, rubber tappers, factory workers, peasants, church communities, students, environmentalists, and development experts.

Programs: On native issues, CEDI worked closely with the Union of Indigenous Nations (UNI) to research and publicize the information needed to fight for Indian rights at the 1988 Constitutional Convention. CEDI has also worked with the Council of Geologists investigating the threats to Indian lands from mining operations. And in progress, is production of the ambitious, 18-volume series (*Povos Indígenas do Brasil*) documenting the state of Brazil's native populations.

Special Needs: Financial support for all projects; an information exchange with similar organizations; technical support for cartographic work and using satellite images of deforestation.

Resources: The weekly news bulletin *Aconteceu (Events)*, a monthly magazine *Tempo e Presença (Time & Place)* whose Aug./Sept. 1989 issue was on Amazonia; the document series *Povos Indígenas do Brasil*—four of the 18 volumes are now available; a document series *Terras Indígenas* produced with the National Museum on the Indian lands in Amazonia (in Portuguese with English abstracts); the report *Mineração em Terras Indígenas no Brasil*, on mining claims on Indian lands; an extensive catalogue of publications, texts, newspaper and journal clippings, videos, photos about Brazil's Indians.

CEDI is now publishing a biweekly analysis of political and economic news in Brazil via Econet.

Contact: Tony Gross
CEDI
Av. Higienópolis 983
01.238 São Paulo, SP
Brazil
Phone: (11) 825-5544
Fax: (11) 825-7861
GEO2: CEDI
Telex: 2137982
Econet: RAN:TROPICTIMBER

Can cattle pasture be returned to ecologically sound uses? Photo: ©Aguirre/Switkes

▶ **Centro de Estudos e Atividades de Conservação da Natureza (CEACON)**

(Center of Studies and Activities for the Conservation of Nature)

Objectives: CEACON builds public and government support for conservation and appropriate development.

Programs: CEACON's Project Amazonia studies development impacts in Pará state. CEACON also has campaigns to fight the use of agricultural chemicals and to establish environmental education programs.

Special Needs: Funding for the training of environmental teachers and for logging onto a computer information network.

Resources: A monthly newsletter.

Contact: David Stevens
Technical Coordinator
CEACON
C.P. 20684
01.498 São Paulo, SP
Brazil
Phone: (11) 881-8422
Fax: (11) 372-800

▶ **Centro de Filmagens Ambientais**

(Center of Environmental Filming)

Objectives: The Center was founded by the well-known activist Vanderley de Castro to protect the environment and support grassroots movements in Amazonia by producing TV films from the movements' perspectives. The Center also helps organize development projects.

Programs: In addition to producing films, the Center has played a key role in organizing public forums, such as the *Semana da Paz* (Week of Peace) in Goiânia on land conflicts, agrarian reform, and the Amazon; in organizing the Center for Training Rubber Tappers in Acre and the Center for Indian Training for Resource Management in Goiânia; and in developing a pilot project with GAIA and the landless in Pará to convert deforested cattle pasture into ecologically sound farms.

Resources: The Center distributes the *Decade of Destruction*, a ten-part film series on environmental destruction in the Amazon. (Includes the segments *Banking on Disaster* and *On the Trail of the Uru-Eu-Wau-Wau*).

Contact: Vanderley Pereira de Castro
Centro de Filmagens Ambientais
Universidade Católica de Goiás
Av. Universitária, 1440
Setor Universitário
74.000 Goiânia, GO
Brazil
Phone: (62) 225-1188, ext. 170

▶ **Centro de Trabalhadores da Amazônia (CTA)**

(Workers' Center of Amazonia)

Objectives: CTA is comprised of union organizers who act as an advisory support group for rubber tappers in the Rural Workers Union of Xapuri, Acre. CTA workers have organized a rubber tappers' grassroots education program, *Projeto Seringueiro* (Project Rubber Tapper), which has been responsible for building 19 schools in the Xapuri area and training more than 50 rubber tappers as teachers.

Special Needs: Funding for new schools.

Contact: Gumercindo Rodrigues
CTA
Av. Nações Unidas
1538 - sala 2
69.900 Rio Branco, AC
Brazil
Phone: (68) 226-3701

▶ **Centro de Trabalho Indigenista (CTI)**

(Indigenist Support Center)

Objectives: CTI was founded in 1979 by anthropologists to support Indian struggles for autonomy.

Programs: CTI is working on the demarcation of lands of the Guaraní–Mbya, Waiãpi, Apinayé, Kricati tribes, and Nambiquara; alternative development projects with the Terena, Krahô, Waiãpi, and Aikewar; health and education programs with the Terena, Kadiweu, Waiãpi, and Krahô; and on producing a "video in the villages" program with the Kayapó-Xikrin, Salumã, and Gavião.

Special Needs: Financial assistance.

Resources: Educational videos produced by CTI and the Indians, including *Pra Ser Krahô (To Be Krahô)*.

Contact: Virginia Valadão
CTI
Rua Fidalga 548 s/13
05.432 São Paulo, SP
Brazil
Phone: (11) 813-3450
Fax: (11) 825-7861

▶ **Comissão pela Criação do Parque Yanomami (CCPY)**

(Commission for the Creation of a Yanomami Park)

Objectives: CCPY is a nonprofit organization founded in 1978 to lead an international campaign for the creation of a Indian park for the Yanomami, the largest traditional Indian group in the Americas. The 20,000 Yanomami are now threatened with the invasion of 45,000 gold miners into their territories, who are polluting their rivers and spreading deadly diseases. They also face the impacts of a recent government decree that has opened up 70 percent of the Yanomami land base to colonization.

Programs: CCPY publicizes the plight of the Yanomami and is pursuing legal strategies to help Yanomami retain their lands. The CCPY was sending in medical teams to treat the Yanomami, to make up for complete neglect by the government agencies. But in 1988, the military expelled CCPY and the medical teams from the area to avoid public scrutiny.

Special Needs: Financial assistance; computers and technical assistance to get timely information circulated internationally on the Yanomami situation.

Resources: The monthly bulletin *Urihi* (in Portuguese and English); action alerts; *Roraima: O Aviso da Morte (Roraima: The Death Sentence)*, published as part of *Ação pela Cidadania* (Citizen Action), a coalition of church, union, scientific, and human-rights organizations working to help the Yanomami; the booklet *Genocídio do Yanomami: Morte do Brasil (Genocide of the Yanomami: Death of Brazil)*; and the film *Povo da Lua, Povo do Sangue (People of the Moon, People of Blood)*.

Contact: Claudia Andujar
Coordenadora
CCPY
Rua Manoel da Nobrega
111, 3° andar, cj. 32
04.001 São Paulo, SP
Brazil
Phone: (11) 289-1200 or 284-6997
Telex: 1126561 ECUMBR

Machadão, a member of the Yanomami group. Photo: Geoffrey O'Connor

▶ **Comissão Pastoral da Terra (CPT)**

(Pastoral Land Commission)

Objectives: CPT was created in 1975, under the military dictatorship, by the liberal wing of the Catholic Church to work for social justice. As Brazil's leading rural worker support group, CPT's main goals are agrarian reform and to protect farmers and squatters from violence. According to Father Ricardo Rezende, who heads up CPT's activities in the south of Pará state, "Violence here is on the rise, not only violent deaths and murders, but torture also. It is not possible in Latin America, which has suffered so much, to speak of God if we are not also capable of creating a more fraternal world through agrarian reform."

Programs: CPT assists in union organizing and the development of sustainable agriculture; documents land conflicts, rural violence, deforestation and its impacts on rubber tappers.

Special Needs: Financial support for emergency situations, such as legal assistance for the Chico Mendes murder case and for sustainable development projects.

Resources: Reports on rural violence and deforestation.

Contact: Presidente
CPT Nacional
Rua 20, No. 251
Centro
74.000 Goiânia, GO
Brazil
Phone: (62) 223-4039

▶ **CPT Regional Offices:**

Contact: CPT-Acre
Márcio Rogério Dagnoni
Secretário Executivo
Praça da Catedral s/n
Palácio do Bispo
Caixa Postal 522
69.900 Rio Branco, AC
Brazil
Phone: (68) 224-2193 or 224-4555
Telex: 682594

Contact: CPT-Amazonas/Roraima
Rua Tapajós, 54-Centro
Cx. Postal 369
69.010 Manaus, AM
Brazil
Phone: (92) 233-0322

Contact: CPT-Araguaia
Padre Ricardo Rezende
Centro Social
68.540 Conceição
do Araguaia, PA
Brazil

Contact: CPT-Marabá
Caixa Postal 80
68.500 Marabá, PA
Brazil

Contact: CPT - Amapá
Av. Procópio Rola, 2719
Caixa Postal 12
68.900 Macapá, AP
Brazil

Comissão Pró-Indio (CPI)
(Pro-Indian Commission)

Objectives: One of Brazil's most reputable Indian support groups, CPI was founded in 1978 when the government was trying to "emancipate" Indians by assimilating them and confiscating their land. CPI members are anthropologists, students, lawyers, and doctors who publicize the struggles of Indian peoples and provide legal assistance and research capabilities to the Indian movement.

Programs: CPI monitors changes in all Indian policies, and has been conducting an in-depth analysis on topics, such the impacts on indigenous peoples of Altamira hydro dams, the Energy Sector's Plan 2010, mining operations, and the Brazilian educational system.

Special Needs: Information on foreign investments in Brazil's energy sector and in other projects affecting native areas.

Resources: The monthly *Boletim Jurídico (Legal Affairs Bulletin)*; a booklet on Cachoeira Porteira Dam; books (in Portuguese) *As Hidrelétricas do Rio Xingu e as Populações Indígenas (The Hydroelectric Dams of the Xingu River and Indigenous Populations)*, (now being translated into English by Cultural Survival); and *A Questão da Mineração em Terra Indígena (The Question of Mining on Indian Land)*; and *A Questão Indígena na Sala de Aula (The Indigenous Question in the Classroom)*. CPI also has a resource library of videos and films.

Contact: Lúcia Andrade
Comissão Pró-Indio
Rua Ministro Godoy 1484
05.015 São Paulo, SP
Brazil
Phone: (11) 864-1180
Telex: 11-82153

Comissão Pró-Indio Acre (CPI-AC)
(Pro-Indian Commission of Acre)

Programs: As part of its education work, CPI-AC has helped train 37 Indian teachers and publish bicultural educational textbooks. In the health field, CPI-AC has trained 24 native people as health workers, assisted with vaccinations, and with the preparation of educational materials on health. CPI has helped Indians set up cooperatives to market rubber and traditional handicrafts. CPI is now working on a video documentation project to be used both for teaching and for cultural documentation.

Special Needs: Financial and technical support to continue research projects in biology, ethnobotany, and psycho-social linguistics, and Indian history in the Americas. CPI-AC also wants to forge closer links with support groups around the world. "The peoples of the forest want allies from throughout the world, not only to denounce problems," says Terri Vale de Aquino, co-founder of CPI-Acre, "but to offer concrete resources to help these forest communities. To help the Indians' and the rubber tappers' organizations will do more to save the forest than one thousand articles in *The New York Times*."

Resources: Educational materials, including textbooks prepared by native people of Acre.

Contact: Nietta Monte or
James Cristina
CPI- AC
Rua Rio Grande do Sul s/n
69.900 Rio Branco, AC
Brazil
Phone: (68) 244-1332

Conselho Indigenista Missionário (CIMI)
(Missionary Council for Support of Indigenous Peoples)

Objectives: Founded in 1972 by the liberation theology movement in the Catholic Church, CIMI is comprised of Catholic missionaries who support Indian rights to self-determination. It is the largest and one of the most outspoken Indigenist groups in Brazil today

Programs: CIMI's principal work is in supporting community based education and health projects, and in providing legal and political assistance. In 1988 CIMI actively campaigned for the protection of Indian rights in the new Brazilian constitution and obtained widespread media coverage of Indian problems with development projects, such as the Calha Norte project and oil exploration in the Javari area.

Special Needs: Broader foreign support for Indian rights and information on alternative technologies.

Resources: The excellent newspaper *Porantim* (in Portuguese), 10 times/yr.; *Amerindia*, (in Spanish), 10 times/yr., about Indians throughout the Americas; detailed reports (in English) on the Calha Norte military project and on the colonization of the Javari Valley; audiovisual materials and videos on the Balbina dam (*Balbina, Destruição e Morte*) and the Carajás mining project (*Carajás*); a monthly bulletin, *Informativo Calha Norte (Information on Calha Norte)*.

Contact: Antonio Brand
CIMI
Edifício Venâncio III
Sala 311
C.P. 11-1159
70.084 Brasília, DF
Brazil
Phone: (61) 225-9457
Telex: 614293

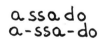

Spelling exercise from CPI-AC's workbook for Indian rubber tappers

CIMI Regional Offices:

Contact: CIMI Norte I
Guenter Francisco
Loebbens, Coordenador
Rua Tapajós
54 Centro
Caixa Postal 984
69.000 Manaus, AM
Brazil
Phone: (92) 233-5020

Contact: CIMI Norte II
C.P. 1454
66.000 Belém, PA
Brazil
Phone: (91) 222-8551

Contact: CIMI Rondônia
C.P. 121
78.900 Porto Velho, RO
Brazil
Phone: (69) 221-9175

Contact: CIMI Acre
Palácio Episcopal
C.P. 473
69.900 Rio Branco, AC
Brazil
Phone: (68) 224-6974

Contact: CIMI Mato Grosso
Rua João Gomes Sobrinho
793—C.P. 884
78.000 Cuiabá, MT
Brazil
Phone: (65) 322-6721

Contact: CIMI Maranhão/Goiás
Casa Paroquial
C.P. 001
65.380 Bom Jardim, MA
Brazil

Coordenação Nacional de Geológos (CONAGE)

(National Coordinating Commitee of Geologists)

Objectives: CONAGE is a progressive geologists' organization that provides Indians with scientific and environmental data related to development projects.

Programs: CONAGE played an important role at the Constitutional Assembly for Indian Organizations by providing information on mining operations and their social and environmental impacts on Indians.

Contact: Wanderlino de Carvalho
CONAGE
Rua 68, n° 727
74.130 Goiânia, GO
Brazil
Phone: (68) 223-7583
Telex: 62-340

Fundação de Apoio a Vida nos Trópicos (ECOTROPICA)

(Foundation of Aid to Life in the Tropics)

Programs: ECOTROPICA is working to protect ecologically sensitive areas in the *pantanal* (wetlands), *cerrado* (savannah), and the Amazon.

Contact: Adalberto S. Eberhard
ECOTROPICA
Av. Rubens de Mendonça
917, sala 502
78.000 Cuiabá, MT
Brazil
Phone: (65) 321-6777
Telex: 65-2361

Fundação Brasileira para a Conservação de Natureza (FBCN)

(Brazilian Foundation for the Conservation of Nature)

Programs: FBCN is an environmental group affiliated with International Union for the Conservation of Nature and active in programs on biodiversity, habitat protection, and environmental education.

Resources: newsletter and brochures.

Contact: Wanderbild Duarte
de Barros
Presidente
FBCN
Rua Miranda Valverde
103 – Botafogo
22.281 Rio de Janeiro, RJ
Brazil
Phone: (21) 226-2654
Telex: 21-37984 FBCN

Fundação Gaia

(Gaia Foundation)

Objectives: Gaia was founded by Brazil's foremost environmentalist José Lutzenberger, who was awarded the Right Livelihood Award (the alternative Nobel Prize) for his innovative work in developing and promoting alternative agriculture. Gaia's goals are to bring an end to ecologically damaging forms of agriculture and to preserve natural and traditional systems, particularly those of native peoples.

Programs: Gaia provides technical assistance on appropriate farming to squatters in Pará State; gives advice on sustainable development of native plants and animals to the Xavante Indians of Mato Grosso; conducts educational programs on the impacts of agricultural chemical use; and is creating botanical gardens and archives on environmental issues.

Special Needs: Financial assistance; an exchange of publications.

Resources: The ten-film series *Decade of Destruction*; books and slideshows on alternative agriculture.

Contact: Ralf Wagner
Consultor
Fundação Gaia
Rua Jacinto Gomes, 39
90.040 Porto Alegre, RS
Brazil
Phone: (512) 313-105
Fax: (512) 802-233

Carajás mining operation. Photo: ©Aguirre/Switkes

▸ **Fundação Pró-Natureza (FUNATURA)**

(Pro-Nature Foundation)

Objectives: FUNATURA is Brazil's principal conservation group, which was founded by Maria Tereza Jorge Pádua, the "Mother of Brazil's National Parks." FUNATURA played a key role in the designations of Brazil's national parks, and the research that helped identify ecologically sensitive areas. FUNATURA is working to strengthen the role of the private sector in protecting biological diversity and to educate Brazilians in conservation issues.

Programs: FUNATURA's staff biologists, forest engineers, economists, and ecologists is now working in partnership with international wildlife groups, such as the World Wildlife Fund and the Nature Conservancy, to establish and manage wildlife sanctuaries.

Resources: The *FUNATURA Bulletin*, 3 times/yr.

Contact: Maria Tereza Jorge Pádua
FUNATURA
Cx. Postal 02-0186
70.001 Brasília, DF
Brazil
Phone: (61) 274-5449
Fax: (61) 274-5324

Samaúma tree, Acre. Photos: ©Aguirre/Switkes

▸ **Fundação SOS-AMAZONIA**

(SOS-AMAZONIA Foundation)

Programs: SOS programs focus on environmental education, recycling, and environmental litigation in the urban areas of Acre.

Special needs: Funding, information on recycling technology, and assistance on evaluation of environmental impact reports.

Contact: José Antonio Scarcello
SOS-AMAZONIA
Rua Visconde de Mauá, 213
69.900 Rio Branco, AC
Brazil
Phone: (68) 224-7271

▸ **Instituto de Antropología e Meio Ambiente (IAMA)**

(Institute of Anthropology and the Environment)

Objectives: IAMA is a research and activist institution whose work emphasizes the relationship between environmental problems and the interests of forest peoples. IAMA works mainly with Indian peoples, rubber tappers, riverine populations, and coastal populations on sustainable resource management issues.

Programs: Projects are underway to create a socio-environmental zoning plan for Indians and rubber tappers in the Guaporé region of Rondônia; strengthen legislation on Indian rights; and develop a documentation center.

Special needs: Funding, equipment, and contacts with similar groups.

Resources: *IAMA* magazine (forthcoming); reports on the Balbina dam's impacts.

Contact: Mauro Leonel
Presidente
IAMA
Rua Turi, 16
05.443 São Paulo, SP
Brazil
Phone: (11) 210-1338 or 210-1301
Fax: (11) 257-6455

▸ **Instituto Apoio Jurídico Popular**

(Institute of Assistance for Popular Law)

Projects: The Institute provides legal assistance to popular movements in Brazil such as those fighting the Carajás mining project.

Special needs: Financial support and contacts with similar organizations.

Resources: A report, *Inquérito Civil Programa Grande Carajás (Civil Inquiry into the Greater Carajás Program)*.

Contact: T. Miguel Pressburger or Daniel Reich
Instituto Apoio Jurídico Popular
Avenida Beira Mar 216
sala 401
20.021 Rio de Janeiro, RJ
Brazil
Phone: (21) 262-3406

José Lutzenberger, founder of Gaia Foundation and now Brazil's Environmental Secretary, at the 1986 World Bank meeting.

Mary Alegretti, president of IEA, won the UN's Global 500 Award in 1990.
Photo: MacArthur Foundation

▶ Instituto de Estudos Amazônicos (IEA)

(Institute of Amazon Studies)

Objectives: IEA is an organization of social scientists, lawyers, biologists and other professionals who work to protect the rainforest and the people who live in it. IEA is best known for its key role in developing the concept of the extractive reserve with the rubber tappers of Acre.

Programs: IEA worked closely with the National Council of Rubber Tappers (CNS) and environmental groups around the world to gain international support for extractive reserves. In the Amazon, IEA is backing CNS by training technical teams and local leaders in education, health, and the production and marketing of forest goods. IEA is also a clearinghouse on information related to sustainable development in the Amazon. IEA conducts environmental education workshops at the Chico Mendes Park in Curitiba, Paraná.

Special Needs: Financial support; collaboration with development experts and environmentalists; and information on markets for extractive products and agroforestry systems.

Resources: The newsletter *Boletim Informativo*; a weekly newspaper clipping service; an extensive document and bibliographic file; videos, photographs, and slides on extractive reserves and socio-economic and environmental problems in Amazonia.

Contact: Mary Helena Alegretti
Presidente
IEA
Rua Monte Castel, 380
82.500 Tarumã, Paraná
Phone: (41) 262-9494
Fax: (41) 233-0384
Telex: 55-41-0775

▶ Instituto de Estudos Sócio-Econômicos (INESC)

(Institute of Socio-Economic Studies)

Programs: INESC is a research and lobbying organization on agrarian reform, Indian rights, human rights, and the foreign debt.

Contact: Maria José Jaime
INESC
Sup. Venâncio 2000, Bl. B
50, s435/437
70.333 Brasília, DF
Brazil
Phone: (61) 226-8093
Telex: (61) 4837

▶ Movimento Ecológico Marirí

(Marirí Ecological Movement)

Objectives: Marirí was created to increase ecological awareness and activism in the Amazon state of Acre, by encouraging the public to work closer with the indigenous and rubber tappers communities.

Programs: Marirí is campaigning to stop indiscriminate use of chemical defoliants by cattle ranchers and is researching the medicinal value of plants from the Acrean rainforest.

Special Needs: Funding for the research, publication, and distribution of the medicinal plant's project.

Resources: Reports on field research with Indian groups in Acre and Rondônia.

Contact: Maria de Fátima H. Almeida
Presidente
Marirí
C. Postal 126
69.900 Rio Branco, AC
Brazil
Phone: (68) 224-5473

▶ Movimento de Apoio para a Resistênca Waimiri-Atroari (MAREWA)

(Movement of Support for the Resistance of the Waimiri-Atroari)

Objectives: MAREWA is an organization of church workers, Indians, and native rights activists, who have been campaigning to protect the human rights of the Waimiri-Atroari, an endangered tribe of several hundred Indians who have lost their lands to the Balbina Dam and tin mining operations.

Resources: The booklets *Balbina: Ameaça e Destruição na Amazônia (Balbina: Threat and Destruction in the Amazon)* and *Resistência Waimiri/Atoari (The Resistance of the Waimiri Atoari)*.

Contact: Egydio Schwade
MAREWA
BR-174, Km. 107
Presidente Figueiredo, AM
Brazil

▶ **Movimento Nacional de Defesa dos Direitos Humanos (MNDDH)**
(National Movement in Defense of Human Rights)

Objectives: MNDDH was created in 1982 by the First National Conference of Human Rights to coordinate national campaigns and lend support to the 300 regional and local organizations that fight for human rights in Brazil.

Programs: MNDDH conducts training for human rights activists.

Special Needs: Financial aid and information on human rights campaign stategies.

Resources: A journal on the foreign debt *Análise (Analysis)*, and *Jornal dos Direitos Humanos (Human Rights Report)*.

Contact: Augustino Veit
Secretário Executivo
MNDDH
Setor de Diversões Sul
Ed. Venancio VI, sala 109
74.302 Brasília, DF
Brazil
Phone: (61) 321-4751

▶ **Núcleo de Direito Indígena (NDI)**
(Nucleus of Indigenous Rights)

Objectives: NDI is a lawyers' group founded in the mid-1980s by former federal deputy Márcio Santilli to lobby the Constitutional Assembly on Indian rights. NDI now works closely with the Union of Indigenous Nations to choose legal cases that could set legal precedents for Indian rights.

Programs: NDI monitors policies of the National Congress, the judiciary, and the government agencies on indigenous rights and trains legal workers in Indian law.

Special Needs: Financial support and contact with similar organizations abroad.

Contact: Márcio Santilli
NDI
SQS-106, Bl. A, Apto. 102
70.345 Brasilia, DF
Brazil
Phone: (61) 243-4814

▶ **Operação Anchieta (OPAN)**
(Operation Anchieta)

Objectives: OPAN works to create self-reliance in Indian communities by training health, education, and development experts to transfer their skills to native groups.

Programs: OPAN's training programs are nine months long, including four months in the field, where trainees learn about the development needs of the village and how best to meet those goals.

Special Needs: Financial, information exchange.

Resources:: Teaching materials and dictionaries prepared by OPAN in several Indian languages; reports on OPAN's conferences on Indian education; an information center with a photographic and audio-visual archive; videos on Indians of Javari region; the book *Ação Indigenista como Ação Política (Indian Support as Political Action)*.

Contact: Ivo Shroeder
C.P. 615
78.000 Cuiabá, MT
Brazil
Phone: (65) 322-2980
Telex: 65-2212

▶ **União Protetora do Ambiente Natural (UPAN)**
(Union to Protect the Natural Environment)

Objectives: UPAN is an environmental lobbyist organization dedicated to cleaning up Brazil's polluted rivers and opposing large dam projects.

Programs: UPAN has been monitoring industrial pollution levels in Brazilian rivers and pressuring the government to enforce anti-pollution laws with tanneries and other industries. UPAN is also campaigning to stop mercury contamination of the rivers from gold mining. As part of the International Rivers Network (see IRN, page 69) UPAN is building a Latin American rivers network that would improve communication and cooperation between river protection activists throughout South America.

Resources: The bimonthly newsletter *Sinal Verde (Green Light)* in Portuguese and English; special issue on gold mining, *Red Light for the Illegal Garimpo of Gold*.

Contact: Carlos Cardoso Aveline
Presidente
UPAN
Caixa Postal 189
93.001 São Leopoldo, RS
Brazil
Phone: (512) 927-933
Telex: 511219 XPSL BR

Gold miner at Serra Pelada in Pará.
Photos: ©Aguirre/Switkes

▸ Resource Groups

▸ **Agência Ecumênica de Notícias (AGEN)**
(Ecumenical News Agency)

AGEN is an alternative news network, which publishes information on social, political, and human rights issues from more than 100 church, labor, Indian, and peasant groups.

Special Needs: Funding for operations and news stories.

Resources: *AGEN*, a weekly newspaper, daily bulletins, available by mail or by telex; a monthly digest (in English).

Contact: José Domingos Bragheto
AGEN
Av. Ipiranga 1267, 8° andar
01.039 São Paulo, SP
Brazil
Phone: (11) 229-6734
Telex: 11-25824 AECN-BR

▸ **Associação Brasileira de Reforma Agrária (ABRA)**
(Brazilian Association for Agrarian Reform)

ABRA was created in 1967 by social scientists to research and publish information and studies documenting the need for land reform.

ABRA coordinates events, programs, and communication between unions, universities, popular movements, political parties, government authorities, and international organizations. ABRA is also participating in the rewriting of the state constitutions' environmental platforms.

Special Needs: Financial support

Resources: *Revista Reforma Agrária (Agrarian Reform Magazine)*, published four times/yr. since 1971, is Brazil's oldest journal on agrarian reform; a library on agrarian issues in Brazil, with a clippings file and a data bank of legal and socio-economic information.

Contact: Luis C. Guedes Pinto
Diretor de Coordenação e Supervisão
ABRA
Rua Cândido Gomide, 333
13.070 Campinas, SP
Brazil
Phone: (192) 42-6590

Landless migrate to Rondônia in search of fertile land and a new life but find neither.
Photos: ©Aguirre/Switkes

▸ **Centro de Educação, Pesquisa e Assessoria Sindical e Popular (CEPASP)**
(Center of Education, Research, and Advice for Unions and Popular Movements)

CEPASP provides information and advice to unions and public-interest organizations on organizing, education, and environmental issues, such as the Carajás mining project.

Special Needs: Information on investments in the Greater Carajás Mining Project.

Resources: A monthly bulletin, other periodicals, and a video (forthcoming) on Carajás metal plants.

Contact: Raimundo Gomes da Cruz Neto
Presidente
CEPASP
R. Barão do Rio Branco, 722
68.500 Marabá, PA
Brazil
Phone: (91) 321-1564 or 324-1615

▸ **Centro Mari de Educação Indígena (CMEI)**
(Mari Center of Indigenous Education)

CMEI is a new organization working to improve the quality of education in indigenous areas, and to publish materials for the general public about the problems native people face. CMEI wants to expand contacts with other groups involved in Indian education.

Contact: Mariana K.L. Ferreira
CMEI
Depto. de Antropología,
Universidade de São Paulo
FFLCH
Cx. Postal 8105
05.508 São Paulo, SP
Brazil
Phone: (11) 210-2122 or 263-3024 ext. 475 or 574

▸ **Centro de Pastoral Vergueiro (CPV)**
(Vergueiro Church Center)

CPV publishes a collection of news, editorials, and features from the national and alternative press on social and political issues.

Resources: The bi-monthly magazine *Quinzena*.

Contact: CPV
Caixa Postal 42.761
04.299 São Paulo, SP
Brazil
Phone: (11) 273-6533 or 273-9322

▸ **Coletivo de Realizadores de Audiovisuais da Amazônia (CRAVA)**
(Collective of Audiovisual Artists of the Amazon)

CRAVA is the only independent filmmakers' group in the Brazilian Amazon and is responsible for producing many films on culture and social change.

Contact: Januário Guedes
CRAVA
Museu da UFPA
Av. Gov. José Malcher 192
66.000 Belém, PA
Brazil
Phone: (91) 224-0871

Esquadrão da Vida Produções Culturais

(Squadron of Life Cultural Productions)

The Squadron of Life is a new organization, which formed to create an information network about the Amazon and other environmental issues.

Resources: The *Jornal Viva Alternativa (Alternative Life Journal)*.

Contact: Ary Pararraios
Jornal Viva Alternativa
Caixa Postal 04/081
70.312 Brasília, DF
Brazil
Phone: (61) 242-6736
Fax: (61) 224-5342

Federação de Orgãos para Assistência Social e Educacional (FASE)

(Federation of Organizations for Social and Educational Assistance)

FASE provides educational materials on social and environmental issues to grassroots organizations.

Resources: The booklet *O Testamento do Homem da Floresta: Chico Mendes por Ele Mesmo (The Testament of the Man of the Rainforest: Chico Mendes in His Own Words)*, a collection of interviews with Chico Mendes that is published in English as *Fight for the Forest*.

Contact: FASE
Rua Bento Lisboa, 58
22.227 Rio de Janeiro, RJ
Brazil
Phone: (21) 285-2998
Telex: 21-34201

Fundação de Tecnologia do Estado do Acre (FUNTAC)

(Technological Foundation of the State of Acre)

FUNTAC started out as an independent group of engineers working on appropriate technology but is now the state agency responsible for regional planning, monitoring deforestation in forest reserves, forest management, and working with rubber tappers on setting up extractive reserves.

Resources: *Boletim Informativo*.

Contact: Gilberto Siqueira
Diretor Presidente
FUNTAC
Av. das Acácias s/n - lote 1
zona A
CP 395
69.900 Rio Branco, AC
Brazil
Phone: (68) 226-2304 or 336-2304
Telex: 68-2508

Instituto Brasileiro de Análises Sociais e Econômicas (IBASE)

(Brazilian Institute of Social and Economic Analyses)

Objectives: IBASE is Brazil's foremost documentation center and clearinghouse for information on social change issues in Brazil: minority rights, economic development, foreign debt, human rights, the environment, judicial reform.

Programs: IBASE produces books, reports, publications analyzing social, political and economic situations in Brazil, as well as audiovisuals and alternative radio programming; and conducts special studies in collaboration with grassroots movements. IBASE is the Brazilian node for Alternex, a new computer network linking grassroots organizations in Brazil and to international communications networks, such as Econet and Geonet.

Special Needs: Financial aid; communication with activists abroad.

Resources: The bimonthly bulletin *Dados do IBASE*; the monthly bulletin *Boletim Políticas Governamentais* (in English and French as *Brazil Information*); *Dossiê Amazônia* reports on Amazonian development; the booklet *Carajás: O Brasil Hipoteca Seu Futuro (Carajás: Brazil Maps Out its Future);* videos, photos, audiovisuals, and radio programs; a digest of Brazilian news transmitted three times a week via Alternex.

Contact: Nubia Gonçalves
Programa de Intercâmbio Internacional
IBASE
Rua Vicente de Souza, 29
Botafogo
22.251 Rio de Janeiro, RJ
Brazil
Phone: (21) 286-0348
Fax: (21) 286-0541
GEO2: IBASE
Econet: IBASE
Alternex: 7241212047902
Telex: 2136466 BASE BR

Instituto do Desenvolvimento Econômico-Social do Pará (IDESP)

(Institute of Socio-Economic Development of Pará State)

IDESP is a governmental agency that publishes excellent periodicals and reports on development issues in Pará state.

Resources: The magazines *Pará Desenvolvimento (Pará Development)* and *Pará Agrária (Rural Pará)*.

Contact: IDESP
Av. Nazaré 871
66.000 Belém, PA
Brazil
Phone: (91) 224-4411

Photo: © Aguirre/Switkes

Instituto Nacional de Pesquisas na Amazônia (INPA)

(National Amazon Research Institute)

INPA conducts research in ecology, botany, anthropology, and forestry. An INPA staff scientist, Dr. Philip Fearnside, has studied extensively the impacts of the Balbina hydroelectric dam and widespread deforestation. Many of the Amazonian research projects of foreign institutions are conducted jointly with INPA, including studies of the Amazon's canopy with the French institute; deforestation and climate change with Britain's Institute of Hydrology; and sustainable forest management with Royal Botanic Gardens (UK).

Resources: A newsletter; reports on the impacts of Balbina dam.

Contact: Herbert Otto Schubert
Director
INPA
Estrada do Aleixo - km.3
Coroado
69.000 Manaus, AM
Brazil
Phone: (92) 236-9050
Telex: 92-2269

Movimento Botucatuense Pró-Vida (MBPV)

(The Pro-Life Movement of Botucatu)

MBPV is a three-year-old organization that provides technical and legal assistance in organic farming to extension workers and peasant farmers—especially those in land-reform areas. MBPV also lectures on ecology in the local schools.

Contact: Brian G. Hoeve
MBPV
Caixa Postal 322
18.600 Botucatu, SP
Brazil
Phone: (149) 22-6453
Fax: (149) 22-3648

Museu Joaquim Caetano da Silva

(Joaquim Caetano da Silva Museum)

The museum researches traditional medicine and raises medicinal plants of the northern state of Amapá.

Contact: Museu Joaquim
Caetano da Silva
Rua Feliciano Coelho, 1509
Bairro de Trem
Macapá, AP
Brazil

Museu Paraense Emilio Goeldi

(Emilio Goeldi Museum of Pará)

The Goeldi Museum conducts research projects in anthropology, botany, ethnobotany, and ecology.

Resources: The newsletter *Destaque Amazônia (Amazon Report)*, 3 times/yr; and the textbook for a Goeldi exhibition on Kayapó resource-management techniques called *A Ciência dos Mebêngôkre: Alternativas Contra a Destruição (Alternatives to Destruction: Science of the Mebêngôkre)*.

Contact: Museu Paraense Emilio Goeldi
Av. Magalhães Barata 376
66.040 Belém, PA
Brazil
Phone: (91) 229-1332 or 225-2036

Núcleo de Altos Estudos Amazônicos (NAEA)

(Center for Advanced Amazon Studies)

As part of the Federal University of Pará State, NAEA analyzes the history and current state of Pará's environment and peoples. NAEA has sponsored debates on critical issues facing Pará, including the Carajás mining project and charcoal industry. NAEA sponsors educational projects targeted at smaller communities in the interior of the state.

Resources: Publications *Cadernos NAEA (NAEA Notebooks)* and *Os Grandes Projetos da Amazônia (Large-Scale Projects of the Amazon)*; documentation center with books, videos, slides.

Contact: Núcleo de Altos Estudos Amazônicos
Universidade Federal do Pará
Campus Universitário
Guamá
66.000 Belém, PA
Brazil
Phone: (91) 229-9754

Sociedade pela Preservação dos Recursos Naturais e Culturais da Amazônia (SORPREN)

(Society for the Preservation of Natural and Cultural Resources of Amazônia)

SORPREN is one of the oldest ecological organizations in the Brazilian Amazon. Founded in 1968, it conducts research on agroforestry and marketable rainforest products, and works with schools on environmental education.

Resources: Bulletins.

Contact: Camilo Viana
Presidente
SORPREN
Al. Lúcio Amaral 193
66.040 Belém, PA
Brazil
Phone: (91) 222-1589
Telex: 911967

INPA is working with the Smithsonian Institution and World Wildlife Fund studying wildlife habitats of forest fragments. Photo: R. Bierregaard

SOUTH AMERICA
COLOMBIA

▶ **Indigenous Organizations**

▶ **Consejo Regional Indígena del Vaupés (CRIVA)**

(Regional Indigenous Council of the Vaupés)

Objectives: In 1973, CRIVA was the first regional Indian organization formed in the Colombian Amazon to protect indigenous culture and land rights.

Programs: CRIVA works with 24 tribes of the Vaupés Indian Reserve—including the Tukano, Wanano, Desano, and Cubao—to protect the area from the invasion of more than 10,000 gold panners and coca growers.

Special Needs: Funding for surveillance of the Vaupés.

Contact: Alfonso Gonzáles
Presidente
CRIVA
Mitú, Vaupés
Colombia

▶ **Organización Nacional Indígena de Colombia (ONIC)**

(National Indigenous Organization of Colombia)

Objectives: ONIC formed in 1982 to bring unity to Indian struggles for land, culture, and autonomy. ONIC's activities began mainly in the Andean region and in the last few years expanded throughout the Amazon to include the Amazonian regional organizations Cabildo Indígena Mayor Del Trapecio Amazónico, Confederación Indígena del Alto Amazonas, Confederación Indígena del Medio Amazonas, Consejo Regional Indígena del Guainía, Consejo Regional Indígena del Guaviare, Consejo Regional Indígena del Medio Amazonas, Consejo Regional Indígena del Vaupés, Consejo Regional Indígena Orteguaza Medio, Musurunacuna, Organización Huitoto del Caquetá Amazonas y Putumayo, Organización Indígena del Sur de Colombia, Organización Zonal Indígena del Putumayo, and Unión de Indígenas del Guaviare.

Programs: ONIC's principal programs support land protection; community development; bilingual and bicultural education; health centers incorporating western and traditional medicine; and organizing and training men and women on legal and social welfare issues. ONIC is also working on "500 Years: The Self-Discovery of Our America," a campaign to present the Indian perspective of Colombus's "discovery" of America during official 500-year anniversary celebrations in 1992. ONIC is working closely on the campaign with peasants associations, unions, students, artists, teachers, writers, film makers, and religious groups.

Special Needs: Funding for the organizational programs.

Tikuna Indian in Colombia.
Photo: Francois Correa

Resources: The monthly newspaper *Unidad Indígena (Indigenous Unity)*, and *Boletín Ejectivo (Executive Bulletin)*; videos and books.

Contact: Emilio Fiagama
Secretario
ONIC
Cra. 3a # 15-48
A.A. 32395
Bogotá, DE
Colombia
Phone: (1) 242-8017 or 284-8196

▶ **Farmers**

▶ **Federación Nacional Sindical Unitaria Agropecuaria (FENSUAGRO-CUT)**

(National United Agricultural Labor Federation)

Objectives: The Federation was formed by the merging of rural workers unions with the Central Union (CUT) to fight for agrarian reform and against the alarming increase in political oppression in Colombia. "In our country," say Cruz Emilia Rangel and Luis Carlos Acero, spokesmen of FENSUAGRO-CUT, "the 'dirty war' is marked by the double-edged sword wielded by our government. On the one hand it is fighting the drug traffickers, but it is also issuing decrees making it possible to arrest popular leaders and to hold them *incommunicado* for seven days. These and other measures are designed to impede popular action in clear violation of our human rights." FENSUAGRO-CUT has representation in all of Columbia.

Programs: FENSUAGRO-CUT unionizes workers, creates community-run schools; conducts work programs for rural women; and organizes cooperatives, regional development programs, and community forums on labor, human rights, and development issues.

Special Needs: Funding for training teachers, and for forming cooperatives and small businesses.

Resources: The bimonthly bulletin *Unidad Agraria*; textbooks.

Contact: José Gálviz
Secretário Relaciones Internacionales
FENSUAGRO-CUT
Calle 17 no. 10-16, Of. 104
A.A. 40606
Bogotá, DE
Colombia
Phone: (1) 282-7020 or 281-9968

Support Groups

Comisión Andina de Juristas Seccional Colombiana
(Andean Judges' Commission)

Objectives: The Commission fights for the protection of human rights by bringing Colombian cases to international tribunals and forums.

Programs: The Commission is investigating cases of illegal detention and disappearances during the 'dirty war' and bringing them to the Inter-American Human Rights Commission of the UN. The Commission also sponsors an annual seminar on the use of international mechanisms to defend human rights.

Special Needs: Funding for investigations and legal work; for the annual seminar on international law and human rights; and for publishing.

Resources: The journal *Informativo Legislativo y Jurisprudencial (Legislative and Legal Information)* four times/yr; a bulletin *Justicia y Violencia en Colombia (Justice and Violence in Colombia)*; the proceedings of the seminar on human rights.

Contact: Gustavo Gallón Giraldo
Director
Comisión Andina de Juristas
Seccional Colombiana
Carrera 7
No. 39-34 Oficina 701
Apartado Aéreo 58533
Bogotá, DE
Colombia
Phone: (1) 245-8961 or 245-9138
Fax: (1) 211-2790 or 218-6742

Corporación Grupo Ecológico (GEA)
(Ecological Organization)

Objectives: GEA was formed by students of the Javeriana and Los Andes Universities to focus public attention on the social aspects of environmental issues; to develop community-based resource management programs; and to improve environmental legislation.

Programs: GEA is conducting the Conservation of Tropical Humid Forests program, which includes a feasibility study on the creation of international parks in the Amazon and an analysis of the Putumayo Basin and the upper Amazon River.

Resources: An information center.

Contact: Luis Guillermo Baptiste
GEA
Apartado Aéreo 77633
Calle 99 No. 33-42
Bogotá, DE
Colombia
Phone: (1) 218-8885 or 256-1804

Fundación Manoa
(Manoa Foundation)

Objectives: Manoa was created in 1988 to find appropriate scientific and technical solutions to development problems in the Amazon and Orinoco regions.

Programs: MANOA has been working with Amazonian Indian communities and farmers on agroforestry and other development projects. Manoa is also creating a documentation center, which includes the Centro de Imágenes Amazónicas (Center of Amazon Images), a facility that showcases films about the Amazon to stimulate debate on the region's future.

Special Needs: Financial support for its development programs and film center.

Resources: Report on 1989 conference of anthropologists working in the Colombian Amazon

Contact: Mariano Useche
Director Ejecutivo
Fundación Manoa
Apartado Aéreo 45029
Bogotá, DE
Colombia
Phone: (1) 284-8739

Fundación Puerto Rastrojo
(Puerto Rastrojo Foundation)

Objectives: A small group of Colombian biologists and anthropologists formed the project in 1983 to develop strategies for natural resource management on the Caquetá River in the Amazon, emphasizing the need for preserving biodiversity and the traditional cultures.

Programs: At its biological station on the Caquetá River, the Foundation is studying regeneration of disturbed areas and wildlife management; evaluating land-use patterns by the Indians and settlers and helping them determine their own needs and strategies for economic development; and working with those communities to protect the giant river turtle.

Special Needs: Funding for research and teaching materials; information exchange with other documentation centers; contact with tropical ecology groups; office equipment (computers, photocopier, fax, modem, etc.).

Resources: A documentation center on the Colombian Amazon containing some 3,000 volumes and photos.

Contact: Thomas Walschburger
Director
Fundación Puerto Rastrojo
Apartado Aéreo 241438
Bogotá, DE
Colombia
Phone: (1) 284-9010

Resource Groups

Centro Interdisciplinario de Estudios Regionales (CIDER)
(Center For Research and Regional Development)

Programs: CIDER is a consulting and research group on the planning, administration, and financing of sustainable regional development. CIDER produces a development plan for Guainía; is studying the impacts of mining projects in Colombia; and is teaching environmental education.

Special Needs: Bibliographies on Amazonia and the environment.

Contact: Edgar Forero Pardo
Director
CIDER
Carrera 1 Este N° 18A-10
Bloque B PISO 4°.
Apartado Aéreo 4976
Bogotá, D.E.
Colombia
Phone: (1) 281-4986 or 282-4066
exts.: 2656, 2640, 2641
Fax: (1) 284-1890
Telex: 42343 UNAND CO

A giant Amazonian river turtle.
Photo: ©Alice Levey.

▶ **Corporación Colombiana para la Amazonía Araracuara**

(Araracuara Corporation)

Araracuara conducts research on the use of renewable resources and on integrated development of the Amazonian settlements of Macarena, San José del Guaviare, and Puerto Inírida.

Special Needs: Funding for technical training of the communites.

Resources: The semi-annual journal *Colombia Amazónica (Colombian Amazon)*; the monthly *Boletines Técnicos (Technical Bulletins)*; a documentation center; a herbarium.

Contact: Maria Eugenia Avendaño
Gerente
Corporación Araracuara
Calle 20 # 5-44
Apartado Aéreo 034174
Bogotá, DE
Colombia
Phone: (1) 241-7378 or 282-5802

▶ **Corporación Colombiana de Proyetos Sociales (CORPOS)**

(Colombian Corporation of Social Projects)

Objectives: CORPOS was cofounded in 1982 by Alfredo Molano, Colombia's leading journalist on the Amazon, to promote research and public debate on Amazonian issues: the viability of new settlements; the impacts of development on indigenous communities and the environment; and alternative development strategies. CORPOS is assessing the environmental and social impacts of the proposed Marginal Highway of the Forest (a paved road that will open up the entire Colombian Amazon) and is also analyzing the status of colonization and the Indian areas in the border areas.

Special Needs: Financial support to assess colonization projects in the Negro, Vaupés, and Apaporis river basins, and to publish the findings.

Contact: Alfredo Molano
CORPOS
Cra. 4A, No. 25B-62
Apartado Aéreo N° 59048
Bogotá, DE
Colombia
Phone: (1) 242-8746
Fax: (1) 241-1922

▶ **Corporación de Estudios de Sistemas Ecológicos, Económicos y Sociales (CEES)**

(Corporation of Ecological Systems, Economic and Social Studies)

CEES is a research institution working with CORPOS on natural resource issues, such as the status of colonization and Indian areas of the border regions.

Contact: Julio Carrizosa Umano
Director
CEES
Calle 78 N° 8-32
Bogota, DE
Colombia
Phone: (1) 249-0329

▶ **Corporación Nacional de Investigación y Fomento Forestal (CONIF)**

(National Corporation for Forest Research and Development)

CONIF is a joint Dutch-Colombian project in sustainable forest-resource management.

Resources: the bulletins *CONIF Informa (CONIF News)* and *Serie Técnica (Technical Series)*.

Contact: Dr. Gonzalo de las Salas
Presidente
CONIF
Apartados Aéreos 091676 y 095153
Bogotá, D.E.
Colombia
Phone: (1) 267-6844 or 221-3473
Telex: 213-9219/213-9936 Unicentro or 218-8035 Cra. 15 con 90.

▶ **Núcleo de Estudios de la Amazonía Colombiana**

(Nucleus of Colombian Amazon Research)

The Nucleus conducts research on Indian traditional land use patterns and how they have changed. It holds public forums on social justice, environmental protection, and the recently created *resguardo* (legally recognized communal Indian land) in the Huitoto, Caquetá, and Vaupés areas. The Nucleus also studies ethnobotany and sustainable development, subsoil rights in Amazonia, and the political economy and history of the Colombian Amazon. The Nucleus offers an exchange program for professionals and students.

Special Needs: Contact with similar organizations in Amazonia.

Resources: Research reports.

Contact: Roberto Pineda
Núcleo de Estudios de la Amazonía Colombiana
Universidad de los Andes
Apartado Aéreo 4976
Bogotá, DE
Colombia
Phone: (1) 282-4066

▶ **Núcleo de Estudios Huitoto y Muinane**

(Nucleus of Huitoto and Muinane Research)

The Nucleus conducts research studies of Huitoto and Muinane Indians that will help them preserve and strengthen their culture. The Nucleus is compiling information on language, ethnobotany, and traditional medicine.

Special Needs: Funding for the photographic exhibits; for the publication of children's books; and for radio-program production.

Resources: Books on Amazon culture and nature; the radio program *Amazonia*; 20,000-slide archive on Indian cultures; photographic exhibits for rental: *Amazonía: Naturaleza y Cultura (Amazonia: Nature and Culture)*, *Región del Río Inírida (Inírida River Region)*, *Arte Rupestre Amazónico (Prehistoric Amazon Art)*, *Las Palabras del Poder–Chamanismo (Words of Power–Shamanism)*, *El Cañon de Araracuara (Araracuara Canyon)*, *Mito y Gesto (Myth and Gesture)*, and *Las Palabras de la Coca (Words of Coca)*.

Contact: Fernando Urbina, Director
Núcleo de Estudios
Huitoto y Muinane
Centro de Estudios Sociales
Facultad de Ciencias Humanas
Universidad Nacional
Bogotá, DE
Colombia
Phone: (1) 268-2225 or 263-1619

▶ **Project River Dolphin**

A group of Colombian biologists have begun assessing the condition of river dolphin populations in order to campaign for their legal protection. The project includes an investigation of the black-market trade for dolphin organs and a community education program on dolphin protection.

Special Needs: Project funding.

Contact: Fernando Trujillo Gonzalez
Project River Dolphin
A.A. 13001
Bogot, D.E.
Colombia

SOUTH AMERICA
ECUADOR

▸ **Indigenous Organizations**

▸ **Confederación de Nacionalidades Indígenas de la Amazonia Ecuatoriana (CONFENIAE)**

(Confederation of Indian Nationalities of the Ecuadorian Amazon)

Objectives: CONFENIAE was formed in 1980 to protect Indian territories in the Amazon, promote indigenous unity, and represent native interests in government-backed development schemes such as petroleum, gas, and palm oil operations. CONFENIAE consists of ten organizations, representing 150,000 indigenous people: (the Cofán people) Asociación de Comunidades Indígenas de la Nacionalidad Cofán; (the Quichua) Federación de Comunas "Unión de Nativos del Amazonía Ecuatoriana", Federación de Organizaciones Indígenas de Napo, Federación de Organizaciones Indígenas de Sucumbios, Organización de Pueblos Indígenas de Pastaza; (the Shuar) Federación de Centros Shuar, Federación Independiente del Pueblo Shuar del Ecuador; (the Siona and Secoya) Organización Indígena Siona y Secoya del Ecuador.

Programs: CONFENIAE is now working with the Huaorani—a largely independent Amazonian tribe—and Ecuadorian and U.S. environmental groups to protect Huaorani lands from oil and gas operations. A priority has been placed on helping the Huaorani build schools and demarcate their lands. CONFENIAE is also promoting sustainable development projects and negotiating with the Ecuadorian government on the structure of bicultural education for all Indians.

Special Needs: Funding and technical support for a tribal communications network and sustainable development projects; international support for the protection of Huaorani lands.

Resources: The monthly newspaper *Amanecer Indio*; an information center on Indian issues, with photographs and slides.

Contact: Luis Vargas
Presidente
CONFENIAE
Casilla 807,
Puyo, Pastaza
Ecuador
Phone: 885-461

Contact: Ampam Karakras
Coordinador
Quito Office
Av. 6 de Diciembre 159
Oficina #408
Casilla 4180
Quito
Ecuador
Phone: (2) 543-973

▸ **Confederación de Nacionalidades Indígenas del Ecuador (CONAIE)**

(Confederation of Indigenous Nationalities of Ecuador)

Objectives: CONAIE is the national indigenous organization of Ecuador, representing CONFENIAE—the Indian confederation of the Amazon—and Indian people throughout Ecuador.

Programs: CONAIE is taking a leading role in organizing the Ecuadorian portion of the "500 Years of Indian Resistance."—a Latin American campaign, which will present the Indian perspective of Colombus's "discovery" of the Americas during the official anniversary celebrations in 1992.

"There is an urgent need to organize an indigenous response to the celebration of the so-called 'discovery' of America, now officially called by Spain, the U.S., and the Vatican as 'The Encounter of Two Worlds,'" says CONAIE. "From our indigenous perspective there was no 'encounter.' On the contrary, it was an armed invasion…which plundered our land and our resources."

Contact: Cristobal Tapuy
Presidente
CONAIE
Av. de los Granados 2553 y
Av. 6 de Diciembre
Casilli Postal 92-C
Quito
Ecuador
Phone: (2) 248-930

Upper left: CONFENIAE meets with the Huaorani to discuss oil and gas development. Above: Huaorani of Dayuno Village.
Photos: ©Aguirre/Switkes.

PUMAREN forest surveying teams.
Photo: N. Irvine

OPIP's Wood Products Center.
Photo: ©Aguirre/Switkes

Alfredo Viteri of OPIP.
Photo: ©Aguirre/Switkes

▶ **Federación de Centros Shuar**
(Shuar Federation)

Objectives: Founded in 1964, the Shuar Federation is one of the oldest and strongest indigenous organizations in Latin America linking 23 regional associations and 263 Shuar communities to fight for land rights and economic self-sufficiency.

Programs: The Shuar were pioneers in the field of bilingual/bicultural education for Indians. The Shuars' programs in cultural issues and primary and secondary school education are broadcast on three different radio channels and reach Shuar communities across Ecuador. The Federation also has development projects in community cattle-breeding, crafts cooperatives, sawmills, and traditional medicine. In cooperation with Fundación Natura, the Federation is now experimenting with the breeding of different rainforest animals as stable sources of food. The Federation also operates a small fleet of single-engine planes and has its own trained surveyors working to demarcate the lands for which the Shuar seek legal ownership.

Contact: Federación de Centros
Shuar
Tarqui 809, apartado 4122
Quito
Ecuador
Phone: (2) 540-264, 547-803

▶ **Federación de Organizaciones Indígenas del Napo (FOIN)**
(Federation of Indian Organizations of Napo)

Objectives: FOIN is an organization of the Quichua people from the Napo River valley working to secure the Quichua's rights to self-determination and their traditional territories.

Programs: In collaboration with Cultural Survival (U.S.), FOIN started Project Letimaren (Land Titling and Resource Managment Project) in 1988 after a newly constructed road opened the Sumaco area within their territory to logging companies and to migrant settlers. Letimaren's goal was to obtain land titles for the Quichua and to establish development projects that relied on the rainforest's survival rather than its destruction. Through Letimaren, FOIN trained a Quichua technical team to study the land-tenure status of 46 communities in the area and to promote the value of forestry preservation and sustainable development. In late 1988, the team formally asked the government for the land titles under dispute.

While the request is being cosidered, FOIN started Project Pumaren (Resource Use and Management Program) to train the technical team in new technologies and resource managment skills, including basic forestry, cartography, and land-use capacity methods. The Quichua were also exposed to other projects underway by various Latin American Indians such as the Kuna (Panama), the Yanesha (Peru) and the Awa (Ecuador). By the time their land titles are granted, the Qucihua will be well on their way toward planning and implementing their own resource-management project.

FOIN also has projects in bilingual/bicultural education; in coordinating native and western health-care systems; and in alternative agriculture.

Special Needs: Funding for increasing Pumaren's technical assistance and for training and regular communication among dispersed communities. Tax-exempt donations earmarked for Pumaren can be made through Cultural Survival (see page 67).

Resources: *Proyecto Letimaren: Preparado Para El Futuro (Project Letimaren: Preparing for the Future)* is a slide/video show in Spanish or English (Contact Cultural Survival for more information).

Contact: Nelson Chimbu
Presidente
FOIN
Casilla Postal 217
Tena, Napo
Ecuador
Phone: (2) 886-288

▶ **Organización de Pueblos Indígenas de Pastaza (OPIP)**
(Organization of Indigenous People of Pastaza)

Objectives: OPIP is an organization of the Quichua Indians from Pastaza Province who have been seriously impacted by oil and gas development, and new colonist settlements. OPIP promotes the unique cultural traditions of the Quichua and the unification of communities throughout the region.

Programs: In 1982, OPIP started an education program in the Quichua language and also the Wood Products Center. The center's activities include selective logging of forest hardwoods, a small sawmill, a furniture factory, a reforestation project, and a nursery that sells seedlings of local plants.

Special Needs: Funding for organizational operations.

Contact: OPIP
A.A.790
Puyo, Pastaza
Ecuador
Phone: 885-461

Support Groups

Acción Ecológica
(Ecological Action)

Programs: EA is an environmental and appropriate-technology group, which is lending technical assistance for the demarcation of Huaorani lands.

Resources: The monthly magazine *Aleph*; a documentation and information center on appropriate technology and the environment.

Contact: Esperanza Martínez
Coordinadora
Acción Ecológica
Avda. 12 de Octubre 344 y
Psje. F. Solano
Casilla 246-C
Quito
Ecuador
Phone: (2) 231-234

Comisión por la Defensa de los Derechos Humanos (CDDH)
(Commission for the Defense of Human Rights)

Programs: CDDH is a network of human rights and education activists that help unions, indigenous communities, and students build non-violent human rights movements.

Resources: The bulletin *Testimonio* (*Testimony*) three times/yr; special reports and textbooks.

Contact: Víctor Hugo Jijón
Coordonador General
CDDH
Edificio Parlamento
Oficina No. 614
Casilla Postal N° 1065
Quito
Ecuador
Phone: (2) 529-328 or 569-340
Telex: 21382 SUTASE ED

Fundación Maquipucuna
(Maquipucuna Foundation)

Programs: MF is a conservation organization working to protect "hotspots" (areas of highly threatened ecological system) such as the Zamora Chinchipe Forest, by developing sustainable use strategies.

Resources: The *Bulletin on Ecodevelopment and Biodiversity in Ecuador*; posters.

Contact: Rodrigo Ontaneda
Director
Fundación Maquipucuna
Baguerizo 238
P.O. Box 167-12
Quito
Ecuador
Phone: (2) 236-166 or 501-083
Fax: (2) 565-319
Telex: 21055 FEJUS ED

Fundación Natura
(Nature Foundation)

Objectives: Fundación is an associate partner of the World Wildlife Fund (WWF) and is working on environmental education and protection of threatened wilderness areas.

Programs: Fundación is working with the Shuar Indian Federation to repopulate endangered native fauna; with the Botanical Gardens in Sucua to protect the river systems; and with local communities to protect the Yasuni and Podocarpus Parks, Cuyabeno Reserve, and Cayambe Coca Ecological Reserve. Funding for the park protection came from the debt-for-nature-swap organized by Fundación, WWF, and the Nature Conservancy.

Resources: The bimonthly *Boletín Natura (Nature Bulletin)*, and *Revista Colibrí (Hummingbird Magazine)*.

Contact: Yolanda Kakbadse
Fundación Natura
América 56534
Voz Andes
Casilla 253
Quito
Ecuador
Phone: (2) 434-449, 249-780
Fax: (2) 434-449
Telex: 21211 Natura Ed

Por La Vida
(For Life)

Objectives: Por La Vida is a new coalition of Ecuador's environmental groups that is working with indigenous groups to protect the Amazonian rainforest from devastating development projects, especially the current boom in oil and gas operations. Por La Vida—whose members include Tierra Viva and Acción Ecológica—is using both lobbying and direct-action campaigns to force the government and multilateral development banks to halt these operations until local communities have been given a voice in the development of their lands, their human rights protected, and environmental guidelines strictly followed.

Programs: In March 1990, Por La Vida staged direct actions in Quito and Cuenca to dramatize the plight of the Amazonian people. The group sent out flyers to the residents of both cities notifying them that their historic central plazas will be destroyed the next day by oil and gas development. At the sites on the following day, Por La Vida erected mock oil rigs and held rallies to publicize the issue. Demonstrators at the Quito rally used the balcony of the government building—which is also the President's residence—to address the crowds. The demonstrations got widespread news coverage and contributed to the government's decision one week later to demarcate the Huaorani's traditional lands—an area that is being systematically destroyed by oil and gas development.

Contact: Esperanza Martínez
c/o Acción Ecológica
Avda. 12 de Octubre 344 y
Psje. F. Solano
Casilla 246-C
Quito
Ecuador
Phone: (2) 248-231 or 249-234

Resource Groups

Centro de Comunicaciones Audiovisuales de la Amazonia (CECAAM)
(Center of Audiovisual Communications in Amazonia)

The Center produces and distributes films about the Indians of Ecuador, such as the Shuar.

Contact: Santiago Fruci
CECAAM
Casilla 252 - Suc. 12
Quito
Ecuador
Phone: (2) 562-633
Fax: (2) 504-620

Centro de Documentación e Información de los Movimientos Sociales del Ecuador (CEDIME)
(Center of Documentation and Information on the Social Movements of Ecuador)

The Center is studying the role of social movements and the state in Amazonia from 1860-1990; distributing a bibliography on Amazonian publications; and researching forest policies and their impacts on the Amazon.

Resources: Textbooks on the history and cultures of Amazonia; audiovisual materials; *Amazonía: Nuestra Tierra (Amazonia: Our Land)* and *Sicuanga*, a newspaper clipping service.

Contact: CEDIME
Tamayo No. 938 y Foch
Quito
Ecuador
Phone: (2) 236-128

▶ Centro de Documentación e Investigación Shuar (CEDISH)

(The Shuar Documentation and Research Center)

The Center is part of the Shuar Bilingual and Intercultural Institute, and provides information and conducts research on the Shuar people.

Resources: An archival collection of audiovisual materials, and bibliographic materials on the Shuar culture and Amazonia.

Contact: CEDISH
Bomboiza (Gualaquiza)
Morona Santiago
Ecuador

▶ Ediciones Abya-Yala and Centro Cultural Abya-Yala

(Abya-Yala Editions and Cultural Center)

The Center has published hundreds of books and periodicals about Indians of the Americas and contemporary Amazon issues.

Resources: Catalogue of publications.

Contact: P. Juan Botasso
Director
Ediciones Abya-Yala
Casilla 8513
Quito
Ecuador
Phone: (2) 562-633

▶ Instituto de Estrategias Agropecuarias (IDEA)

(Institute of Farming Strategies)

The Institute conducts research on economic and natural resource management policies and their impacts on agricultural production.

Contact: Neptalí Bonifaz
Executive Director
IDEA
Bossano 617 y Coronel
Guerrero
Quito
Ecuador
Phone: (2) 245-344

SOUTH AMERICA
GUYANA

▶ Indigenous Organizations

▶ Caribbean Organization of Indigenous Peoples (COIP)

Objectives: COIP is a three-year-old umbrella organization of indigenous groups from four English-speaking countries of the Caribbean: Dominica (Carib people), St. Vincent (Carib), Guyana (Arawaio, Warrou, Carib, WaiWai, Wapishana), and Belize (Maya Mopon, Maya Kekchi, Garifuna). COIP promotes cultural pride and the protection of basic human and civil rights.

Programs: COIP works with the UN Working Group of Indigenous Peoples and the World Council of Indigenous Peoples. COIP also conducts and exchange program of human rights activists.

Resources: Occasional newsletter *Indigi-Notes*.

Contact: Dr. Joseph O. Palacio
Coordinator
COIP
P.O. Box 229
Belize City
Belize
Central America
Fax: (11) 78453

▶ Farmers

▶ National Farmers Organization (NFO)

Objectives: NFO is a 6,000-member organization formed in 1977 to serve and promote the interests of farmers in Guyana, especially small-scale farmers, by using appropriate technology, and improving access to markets and nutrition levels. "The concept of an independent farmers' movement is new to these parts," says NFO, "and farmers funding their own organization seems to be a difficult task. Our work is based on the farmers themselves providing the time to organize and motivate their fellow farmers with some help from facilitators."

Programs: NFO has organized a farmers' cooperative for the milling and marketing of rice—Guyana's staple food and export crop.

Special Needs: Information on sustainable agricultural techniques and alternatives to excessive pesticide use.

Contact: Hafiz Rahman
Secretary
National Farmers
Organization
P.O. Box 10207
General Post Office
Georgetown
Guyana
South America
Phone: (2) 61-054 or 72-845

Shuar bilingual classroom. Photo: ©Aguirre/Switkes

▶ **Support Groups**

▶ **Guyana Human Rights Association (GHRA)**

Objectives: GHRA was formed in 1979 by Guyanan activists to protect human rights.

Programs: GHRA provides information on human rights abuses in Guyana and legal assistance for victims of police brutality and government harassment. GHRA also conducts a malaria eradication campaign and organized a Conference of Amerindian leaders in April–May 1987.

Resources: Proceedings of the Amerindian Conference; reports on human rights abuses in Guyana.

Contact: Bishop Randolph George
co-President
GHRA
Lot 27, Brickdam
P.O. Box 10720
Georgetown
Guyana
South America
Phone: (2) 61-789
Telex: GUYISRA

▶ **Resource Groups**

▶ **Amerindian Research Center (ARC)**

ARC conducts research on Amerindian groups and their movements.

Resources: A newsletter.

Contact: Janette Forte
ARC
University of Guyana
Brickdam, Georgetown
Guyana
South America

Photo: © Aguirre/Switkes

SOUTH AMERICA
PERU

▶ **Indigenous Organizations**

▶ **Asociación Interétnica del Desarollo de la Selva Peruana (AIDESEP)**

(Interethnic Association for the Development of the Peruvian Forest)

Objectives: AIDESEP is a confederation of 21 regional native organizations that formed to protect Indigenous lands and culture in the Amazon. AIDESEP member organizations include: (Achual, Candoshi, Shuar) OSHDEM; (Achuar) ORACH; (Aguaruna & Chayahuita) FECONADIC; (Aguaruna & Huambisa) CAH; (Aguaruna) CHAPI SHIWAG, OAAM, ONAPAA; (Amarakaire, Arasaire, Ese-Eja, Huachipaire, Machiguenga, Piro Sapateri, Toyoeri) FENAMAD; (Ashaninka) ANAP, CECONSEC, OAGP, OIRA; (Bora, Huitoto) FECONAPU; (Bora, Huitoto, Ocaina, Yahuas) FECONA; (Candoshi) FECONACADIC; (Cocamilla) FEDECOCA; (Conibo & Shipibo) FECONAU; (Conibo & Shipibo) FECONBU; (Kiichuaruna & Wangurina) ORKIWAN; (Quichua) FECONABABAN; (Shapra) FESHAM.

Programs: AIDESEP has had significant success in obtaining land demarcation for Indian territories. In the Atalaya region, where Indian territories were being taken over by powerful land owners, Peruvian authorities told AIDESEP that there were no funds for the land-demarcation process. But AIDESEP secured its own funding from the Danish government and hired its own technicians for the project.
AIDESEP also represents Indian interests with government agencies working on bilingual education. It is conducting its own bilingual education, traditional medicine, and university scholarship programs for Indians.

Special Needs: Funding for its programs and for technical consultants.

Resources: The newsletter *Voz Indígena (Indigenous Voice)*; monographs.

Contact: Miqueas Mishari
Presidente
AIDESEP
Av. San Eugenio 981
Lima 13
Perú
Phone: (14) 724-605

▶ **Confederación de Nacionalidades Amazónicas del Perú (CONAP)**

(Confederation of Amazon Nationalities of Peru)

Objectives: CONAP was created in 1987, to preserve the traditional cultures, land, and autonomy of Amazonian nationalities and to promote sustainable development.

Programs: CONAP is now conducting sustainable development projects with Indians living in the Manu National Park area and other regions of the Amazon. CONAP is also campaigning against Shell Oil's natural gas project in the Camisea region by publicizing the project's environmental impacts and by urging the government to repeal laws allowing development of the Amazon.

Special Needs: Funding for the Shell Oil campaign and for programs on sustainable development.

Resources: The bulletin *CONAP;* special reports on the Shell Oil project and other development schemes.

Contact: Aníbal Francisco Coñivo
Presidente
CONAP
Av. Ariosto Matellini, 569
Urb. Matellini Chorillos
Lima
Perú
Phone: (14) 678-839

Yanesha Forestry Cooperative land use planning. Photos: Bob Simeon

A strip-shelter-belt harvesting.

Oxen are used instead of heavy machinery to limit destruction during timber removal.

▶ Federación de Comunidades Nativas del Medio Napo (FECONAMN)

(Federation of Native Communities of the Middle Napo)

Objectives: FECONAMN is an organization of 20 native communities in the Napo region that protects Indian rights and the environment.

Programs: FECONAMN is helping eight communities obtain land titles and is conducting a census of Napo communities that will make it possible for them to vote in national elections.

Special Needs: Funding to transport market products to Iquitos.

Contact: FECONAMN
Apartado 216
Iquitos, Loreto
Perú

▶ Federación de Comunidades Campesinas y Nativas de Loreto (FCCNL)

(Federation of Peasant and Native Communities of Loreto)

Objectives: The Federation was formed by the Quichuaruna, Huitoto, and Yagua Indians, and mestizo riverine communities to fight for land protection, the reform of official agricultural policies, and for human rights.

Programs: FCCNL is training Loreto peasants in resource management, land protection, and human rights.

Special Needs: Funding for transportation and organizational expenses.

Resources: Books and bulletins.

Contact: Catalino Lavi
Director
FCCNL
Calle Moro 246
Iquitos, Loreto
Perú
Phone: 237-634

▶ Yanesha Forestry Cooperative

Objectives: Organized in 1986, the Yanesha Co-op is one of the first forestry projects in South America to help the Indians obtain land titles; protect biologically diverse areas from uncontrolled colonization; and to develop innovative methods of agroforestry compatible with traditional forms of land use.

Programs: With land titles obtained from the government in 1986, funding from the U.S. Agency for International Development, and technical assistance from the Tropical Science Centre (Costa Rica), the Yanesha began using "strip-shelterbelt harvesting." The technique involves clear-cutting long narrow strips of forest that will regenerate in 30-40 years. It is similar to the Yanesha's traditional slash and burn method, but allows for the harvesting of the wood rather than its destruction.

Through its affiliate, Fundación para la Conservación de la Naturaleza (FPCN), World Wildlife Fund is now providing most of the support for training in forestry, operating and repairing equipment, driving, mapping, marketing, and business administration. Backed by their Federation of Yanesha Native Communities (FECONAYA), the Yanesha now own and manage all phases of the project. In 1989, they earned $16,000 in export sales and $31,000 domestically.

Contact: c/o FPCN
Chinchón 858-A
Apartado Postal 18-1393
San Isidro
Lima, 27
Perú
Phone: (14) 422-796 or 408-846
Fax: (14) 422-796 or 410-258
Telex: 25129 Penoving

▶ Support Groups

▶ Asociación Ametra 2001

(Ametra 2001 Association)

Objectives: Ametra is revitalizing the use of traditional medicines to treat the serious illnesses spreading among many Peruvian Indians.

Programs: Ametra works with FENAMAD (Native Federation of the Madre de Diós River and Tributaries) on research, health, and education projects in that area and with other communities in the Pucallpa area.

Contact: Ametra 2001
Casilla 42
Puerto Maldonado
Madre de Diós
Perú

▶ Asociación de Ecología y Conservación

(Association of Ecology and Conservation)

Objectives: The Association is an environmental research and lobbyist organization promoting the protection of biodiversity.

Programs: The Association is conducting a campaign against illegal trafficking in rainforest animals and plants and legislation that allows rainforest destruction.

Resources: A bulletin.

Contact: Tony Luscombe
Vice Presidente
Asociación de Ecología y Conservación
Vanderghen 560-2A
Lima, 27
Perú
Phone: (14) 407-276
Fax: (14) 525-337
Telex: 25684PE

Centro para el Desarrollo de la Amazonía Indígena (CEDIA)
(Center for the Development of Indigenous Amazonia)

Objectives: CEDIA helps Indian communities secure titles to their traditional lands through technical, anthropological, and legal assistance.

Programs: CEDIA is working on land titling projects within the Machiguenga, and with groups living within the Manu National Park area.

Special Needs: Funding for the land titling project.

Contact: Lelis Rivera Chávez
Presidente
CEDIA
Pasaje Bonifacio 166
Urb. Los Rosales de Santa Rosa
La Perla, Callao
Perú

Centro de Investigación y Promoción Amazónica (CIPA)
(Center of Amazonian Research and Promotion)

Objectives: CIPA is a organization of social scientists that provides scientific, technical, and legal assistance to Peruvian Indians on community development and health programs. CIPA was instrumental in the founding of the Confederation of Amazonian Nationalities (CONAP).

Resources: The journal *Temas Amazónicos (Amazon Themes)*, *Extracta* and other documents; and books.

Contact: Ana María Chonati
Responsable C.D.
CIPA
Ave. Ricardo Palma 666 D
Miraflores
Lima, 18
Perú
Phone: (14) 464-823

Fundación Peruana para la Conservación de la Naturaleza (FPCN)
(Peruvian Foundation for the Conservation of Nature)

Objectives: FPCN was formed to protect threatened habitats and biodiversity by formulating conservation strategies and gaining public support through environmental education. FPCN works closely with the U.S. Agency on International Development (AID), Peruvian government agencies, the International Union for the Conservation of Nature (IUCN), Conservation International, The Nature Conservancy, World Wildlife Fund, and the UNESCO-World Heritage Committee.

Programs: FPCN is conducting projects in five national parks on habitat protection including the Yanachaga-Chemillén; developing resource management schemes in Manu Reserve; researching the impacts of protected areas on local communities; and teaching environmental studies; providing technical and financial support to the Yanesha Forestry Cooperative.

Resources: The quarterly newsletter *FPCN al Día*; the reports *Documentos de Conservación (Conservation Documents series)*; maintains a library with slides and photos.

Contact: Alejandro Camino
Director de Desarrollo
FPCN
Chinchón 858-A
Apartado Postal 18-1393
San Isidro
Lima, 27
Perú
Phone: (14) 422-796 or 408-846
Fax: (14) 422-796 or 410-258
Telex: 25129 Penoving

Resource Groups

Centro Amazónico de Antropología y Aplicación Práctica (CAAAP)
(Amazon Center of Anthropology and its Practical Applications)

The Center is part of the Catholic church and conducts research and education projects on Amazonian issues.

Resources: *Amazonia Peruana (Peruvian Amazon)*.

Contact: CAAAP
Avenida González Prada 626
Iquitos, Loreto
Perú
Phone: (94) 615-223

Centro de Datos para la Conservación (CDC)
(Conservation Data Center)

The Center was formed by professors of the Faculty of Forestry Sciences of the Universidad Agraria de la Molina and the Nature Conservancy to supply the government and researchers with scientific data on Peruvian biota and conservation for natural resource management projects. CDC carries a compilation of scientific studies on the Tambopata region.

Resources: The bulletin *Amigo y Guarda (Friend and Protector)*.

Contact: Jorge M. Chávez
Administrador
CDC
Universidad Nacional
Agraria La Molina
Facultad Ciencias Forestales
Apartado 456
Lima
Perú
Phone: (14) 352-035 ext. 234
or 352-260 ext. 234
Fax: (14) 422-796

Centro de Estudios Teológicos Amazónicos (CETA)
(Center of Amazon Theological Studies)

The Center conducts research on indigenous Amazon cultures.

Resources: The weekly bulletins *Shupihui* and *Kanatari*; *Monumento Amazónico (Amazon Monument)*, a collection of historical accounts of the Amazon; a resource list.

Contact: Joaquín García
Director
CETA
Putumayo 355
Iquitos, Loreto
Perú
Phone: (94) 234-253
Fax: (94) 233-190

Centro de Estudios y Promoción de Desarrollo (DESCO)
(Center of Studies and Promotion of Development)

DESCO researches and publishes materials on development issues in Peru.

Resources: The monthly magazine *Qué Hacer (What to Do)* and *Resumen Semanal (Weekly Summary)*.

Contact: Federico Velarde
Presidente
DESCO
León de la Fuente 110
Lima, 17
Perú
Phone: (14) 617-309 or 610-984

▶ **Copal**

COPAL is a small group of anthropologists that has been providing research, legal, and technical assistance to Peru's Amazonian Indians for the last 15 years. COPAL has worked in an advisory capacity for Interethnic Association for the Development of the Peruvian Amazon (AIDESEP).

Resources: The journal *Amazonía Indígena (Indigenous Amazonia);* an information center.

Contact: Alberto Chirif
Copal
c/o CETA
Putamayo 355
Iquitos, Loreto
Perú
Phone: (94) 234-253
Fax: (94) 233-190

▶ **Escuela Amazónica de Pintura Usko-Ayar**

(Usko-Ayar Amazonian School of Painting)

The school was formed by anthropologist Luis Eduardo Luna and painter Pablo Amaringo to provide free art education for children in Pucallpa area.

Programs: The school is creating an ethnobotanical garden.

Contact: Pablo Amaringo
Director
Escuela Amazónica de
Pintura Usko-Ayar
c/o FPLN
Chinchón 858-A
Apartado 18-1393
Lima, 27
Perú
Phone: (14) 422-796

▶ **Instituto de Desarrollo y Medio Ambiente (IDMA)**

(Institute of Development and the Environment)

The Institute researches the impacts of development policies and projects on the environment.

Resources: The bulletin *Medio Ambiente (Environment)*.

Contact: Carlos Herz Sáenz
Director
IDMA
General Suárez 1330
Miraflores
Lima, 18
Perú
Phone: (14) 227-979

The Usko-Ayar school house and painting class.
Photos: ©L. Luna

▶ **Proterra**

(Pro-earth)

Proterra is an environmental law organization that is establishing national networks for environmental protection and supporting sustainable development in the Central Forest.

Resources: A study of Peruvian environmental laws; an information center with library, videos, and photos.

Special Needs: Funding for a four-wheel-drive vehicle and to train and hire forest rangers.

Contact: Antonio Andaluz
President
Proterra
Avenida Esteban Campodónico Nº 208
Urbanización Santa
Catalina, La Victoria
Apartado Postal 2731
Lima, 13
Perú
Phone: (14) 723-800
Telex: 26017 PE AGROCONS

▶ **Sociedad Peruana de Derecho Ambiental (SPDA)**

(Peruvian Society of Environmental Law)

SPDA is a group of lawyers and law students who analyze environmental law and consult organizations and agencies on environmental contamination problems and inappropriate development schemes. SPDA also conducts seminars, classes, symposia, and debates on environmental law.

Resources: A comparative analysis of environmental laws in Peru and Latin America.

Special Needs: Funding for office equipment.

Contact: Jorge Caillaux
Presidente
SPDA
Plaza Jorge Arróspide #9
San Isidro
Lima 27
Perú
Phone: (14) 400-549 or 700-721
Fax: (14) 70311
Telex: 20147 PE DROKASA

SOUTH AMERICA
SURINAME

▶ **Tunasarapa Suriname**

Objectives: Tunasarapa is an Indian support organization-in-exile in The Netherlands.

Programs: Tunasarapa runs an information center and raises funds for humanitarian aid and support for Caraiben, Lokono, Trio, Aloekoejana, Wajarekoeles, and Wamas peoples of Suriname, who have been decimated by the civil war of 1986. Of the 39 indigenous communities (with populations from 75 to 2,000 each), 26 have been totally destroyed.

Contact: George Pierre
Representative
Tunasarapa Suriname
Postbus 10497
100IEL Amsterdam
The Netherlands

▶ **Moiwana**

Objectives: Moiwana is Suriname's principal human rights group.

Programs: Moiwana's office provides information and builds public support for human rights cases.

Contact: Stanley Rench
Director
Moiwana
Surinamestraaht 36
P.O. Box 2008
Paramaribo
Suriname

George Pierre of Tunasarapa Suriname. Photo: ©Aguirre/Switkes

SOUTH AMERICA
VENEZUELA

▶ **Indigenous Organizations**

▶ **Asociación Civil Indígena de los Pueblos Yucpa (ACIPY)**

(Indigenous Civil Association of Yucpa Peoples)

Objectives: ACIPY is an organization of the Yucpa Indians working to protect native lands, cultural integrity, and the right to self-determination.

Contact: Javier Armato
Presidente
ACIPY
Misión los Angeles del Tukuko
Apartado 006 Machiques
Perijá
Zulia
Venezuela

▶ **Consejo Indio de Venezuela (CONIVE)**

(Indian Council of Venezuela)

Objectives: CONIVE formed in 1989 by 21 ethnic groups to promote unity among indigenous people and fight for the protection of lands and human rights.

Programs: CONIVE provides information on the social and economic problems faced by the ethnic groups today in Venezuela.

Special Needs: Funding for training Indian leaders.

Contact: Noelí Pocaterra
Presidenta
CONIVE
Apartado Postal 5156
Caracas 1010
Venezuela
Phone: (2) 545-1754

▶ **Support Groups**

▶ **Centro de Educación, Promoción y Autogestión Indígena (CEPAI)**

(Center for Indigenous Education, Promotions and Self-sufficiency)

Objectives: CEPAI was formed in 1972 by Jesuits, teachers, and laypersons who had experience working with the Yekuana peoples on economic self-sufficiency projects and wanted to expand their program to include more indigneous groups.

Programs: CEPAI has established a storage center in Puerto Ayacucho for products of the cooperatives it helped develop. The center also obtains financial credit and provides technical advice to the communities. The following indigenous cooperatives receive CEPAI's support.

Asociación Piaroa de Productores de Cacao (APIPROCA)

(Piaroa Association of Cocoa Producers)

APIPROCA consists of 15 communities of the Manapiare-Guaviarito Basin and the Alto Suapure for the communal cultivation of cocoa.

Asociación Yekuana de Productores de Cacao (AYEPROCA)

(Yekuana Association of Cocoa Producers)

Six communities of the Alto Ventuari and Alto Orinoco work in AYEPROCA to produce cocoa.

Empresas Arawak del Guainía, Río Negro y Casiquiare (EAGUANCA)

(Arawak Cooperatives of the Guainía, Rio Negro and Casiquiare)

Twelve Curripaco, Guarequena, Baniba, Geral and Baré communities of the Rio Negro produce traditional crafts and the *chiquichiqui* fiber.

Empresa Piaroa de Producción de Miel (EPIAMIEL)

(Piaroa Cooperative for the Production of Honey)

The Piaroas of the Guanay area organized EPIAMIEL to increase income from honey production.

Guahibo Samariapo Miel (GUASAMI)

(Guahibo Samariapo Honey Cooperative)

The Samariapo community of Guahibos, south of Puerto Ayacucho formed GUASAMI to produce honey.

Productores de Aceite de Seje (PUORIBU)

(Producers of Palm Oil)

PUORIBU formed in the Piaroa region of the Alto Suapure to produce banana meal and palm oil, which is used for medicinal purposes.

Sánema Miel Alto Parú (SANEMAP)

(Sánema Honey of Alto Parú)

SANEMAP started as a bee-keeping cooperative in 1981 and has since expanded to include cattle breeding, carpentry, and farm machinery repair, and furniture building.

Shaponos Unidos Yanomami del Alto Ventuari (SUYAO)

(United Yanomami Villages of the Alto Ventuari)

SUYAO produces handicrafts using the collection of the *mamure* vine in the Mavaca, Mavaquita, Platanal, Ocamo and El Sejal Yanomami centers. SUYAO uses the cooperative activities to promote Yanomai unity.

Unión Makiritare del Alto Ventuari (UMAV)

(Makiritare Union of the Alto Ventuari)

UMAV started in the Alto Ventuari region to improve cattle breeding and farming methods of the Yekuana people.

Unión de Sitios Arawak del Medio Orinoco (USAMO)

(Union of Arawak Locations of the Middle Orinoco)

USAMO has unified five communities of Baniva, Curripaco and Guarequena through the producing and marketing of *chiquichiqui* fiber.

CEPAI or any of the cooperatives can be reached through:

Contact: José Luis Echeverría
Director General
CEPAI
Edificio Jesús Obrero
Calle Real de las Flores
de Katía
Apartado Postal 30025
Caracas, 103
Venezuela

Phone: (2) 424-001

▶ Centro de Servicio de la Acción Popular (CESAP)

(Center of Service for Popular Action)

Objectives: CESAP supports self-sufficiency and economic development of the indigenous peoples.

Programs: CESAP has helped create cooperatives for producing and marketing traditional handicrafts.

Contact: Rubén Montoya
Director
CESAP
Puerto Ayacucho
Territorio Federal del
Amazonas
Venezuela

▶ Fundación Venezolana para la Conservación de la Diversidad Biológica (BIOMA)

(Venezuelan Foundation for the Conservation of Biological Diversity)

Objectives: BIOMA is a conservation organization founded to protect sensitive ecosystems and develop resource management programs with government agencies and the private sector.

Projects: BIOMA is developing a management plan for the La Neblina National Park. It has established a Conservation Data Center, one of ten CDCs created in Latin America and the Caribbean with the help of the Nature Conservancy to provide government agencies, development banks, and conservation organizations, with an updated inventory of Venezuela's flora and fauna.

Resources: Videotapes about BIOMA's projects, photos, posters, T-Shirts and maps.

Contact: Aldemaro Romero
Director Ejecutivo
BIOMA
Apartado 1968
Caracas, 1010-A
Venezuela

Phone: (2) 571-8831 or 571-6009

Young Yanomami displaying sardines for sale at the SUYAO cooperative.
Photo: Napoleon A. Chagnon

Programa Venezolano de Educación-Acción en Derechos Humanos (PROVEA)

(Venezuelan Program of Education-Action on Human Rights)

Objectives: PROVEA works to promote both the political and social rights of Venezuelans—their right to self-determination, and their rights to health, housing, education, and employment.

Programs: PROVEA provides training for human rights attorneys; legal action on human-rights violations; preparation of teaching materials for schools; documentation of human-rights abuses.

Resources: Monthly bulletin *Referencias (References)*; a documentation center with newspaper clippings and photos.

Contact: Dianora Contramaestre
Equipo Coordinador
PROVEA
Apartado Postal 5156
Carmelitas
Caracas, 1010-A
Venezuela
Phone: (2) 541-0565 or 541-7717
Fax: (2) 541-7717

Sociedad Conservacionista Audubon de Venezuela

(Audubon Conservation Society of Venezuela)

Objectives: The Society is a research, education, and lobbyist organization working to protect Venezuela's rainforests and coastal wetlands.

Programs: Audubon is completing an analysis of the Venezuelan timber industry; working with the National Parks Institute to train and manage park rangers; establishing scientific research stations in the rainforest; and conducting environmental education programs. Audubon is also conducting "eco-tours" of the Venezuelan wilderness areas.

Resources: A newsletter; reports.

Contact: Juan Antonio Sans Uranga
Presidente
Sociedad Conservacionista
Audubon de Venezuela
Apartado Nº 80450
Caracas, 1080
Venezuela
Phone: (2) 91-38-13

Unuma-Sociedad Civil de Apoyo al Indígena

(Unuma Support Group to Indigenous People)

Objectives: Unuma formed in 1989 to support Indians' efforts to revitalize their cultures, build self-sufficiency, and prevent deterioration of their health.

Programs: Unuma provides legal assistance to Indians on land and constitutional rights issues, and publishes documents for orienting the work of native-rights support groups. Unuma also works closely with the Indian Council of Venezuela (CONIVE) on its leadership training, cooperatives, and medical aid.

Programs: Technical assistance in economic development projects and organizational training.

Contact: Haidée Seíjas
Unuma
Apartado Postal 68356
Caracas, 1040-A
Venezuela
Fax: (2) 927-575 or 919-545
(c/o Dirección General,
Biblioteca Nacional)

Resource Groups

Centro de Estudios de Enfermedades Tropicales Simón Bolívar (CAISET)

(The Simón Bolívar Center for the Study of Tropical Diseases)

Funded in 1985 by biologists and anthropologists to coordinate applied research on tropical diseases.

Contact: Dr. Petra Landa
Directora
Centro de Estudios de
Enfermedades Tropicales
Puerto Ayacucho
Territorio Federal del
Amazonas
Venezuela

Grupo Ecológico Bolívar (GREBO)

(Bolívar Ecological Group)

GREBO trains teachers in environmental education and publicizes the environmental and social impacts of rainforest destruction in Venezuela, especially from gold mining and other industries.

Resources: Articles GREBO has written for the Ecological Page in the newspaper *El Bolivarense*; 17 audiovisual programs for schools; an information center with slides, books, photos, maps, magazines.

Contact: Dr. Luis Rojas
Co-director
GREBO
Apartado Postal 527
Ciudad Bolívar 8001
Estado Bolívar
Venezuela
Phone: (85) 43-857 or 40-361

Proyecto Amazonas

(Amazon Project)

The Project was formed by specialists in medicine, dentistry, agriculture, and architecture to help indigenous populations and peasants on sustainable development projects.

Contact: Proyecto Amazonas
Decanatura de la Facultad
de Medicina
Universidad Central de
Venezuela
Caracas, 1050
Venezuela

Sociedad Conservacionista Aragua

(Aragua Conservation Society)

The Society trains teachers in environmental education and institutes programs for the conservation of Henri Pittier National Park.

Contact: Diógenes A. Hermoso
Coordinador
Sociedad Conservacionista
Aragua
Apartado Postal 5115
El Limón
Maracay 2105-A Aragua
Venezuela
Phone: (43) 831-734

Probe International is investigating the environmental and social impacts of Petro Canada's oil and gas operations in the Ecuadorian Amazon. Photo: Peggy Hallward.

NORTH AMERICA
CANADA

▶ **Probe International**

Objectives: Probe is a 17,000-member citizens' group of development and environmental experts who monitor the impacts of Canada's foreign aid projects and trade policies in the Third World. Probe is the leading Canadian organization devoted to increasing public accountability of the policies and projects of the Canadian International Development Agency (CIDA) and Canadian corporations.

Probe has been working closely with other North American organizations to reform the development policies of multilateral development banks (MDBs) while also working to stop specific MDB projects that are environmentally and socially damaging. Probe also works with groups in developing countries by providing them with technical and scientific information about MDB projects and by building Canadian public support for their campaigns through speaking tours, press releases, conferences, and letter writing campaigns. Probe's nationally syndicated articles are carried by some ten daily and 100 weekly newspapers reaching more than one million Canadians.

Programs: In 1989, Probe played a leading role in building Canadian support for the Kayapó Indians' fight against the Altamira hydroelectric dams. Probe co-sponsored Kayapó speaking tours in Canada and raised funds for the Kayapós' conference in Altamira, Brazil in February 1989. Probe has also lobbied the World Bank to attach greater environmental and social guidelines to the new loan for the Polonoroeste resettlement project in Brazil.

A recent Probe media campaign has focused on Petro-Canada's oil drilling operations in the Ecuadorian Amazon, which include road construction through Huaorani Indian territory that threatens to open up the area to colonization. Probe is also investigating the Amazon activities of Canadian mining companies such as Brascan, Inco, and Alcan.

Resources: *Probe Alert*, an action alert published six times a year.

Contact: Peggy Hallward
Forestry Research Director
Probe International
225 Brunswick Ave.
Toronto, ON M5S 2M6
Canada
Phone: (416) 978-7014
Fax: (416) 978-3824

▶ **Scarboro Foreign Mission Society**

Objectives: The Scarboro Foreign Mission Society is a missionary group that lends international solidarity to overseas affiliates on human rights and social-change issues through its Justice and Peace Office.

Programs: Scarboro missionaries working in the Brazilian Amazon have attracted national news coverage in Canada, focusing on the environmental and social impacts from the Balbina dam, particularly on the Ataoari-Waimiri Indians. Along with other progressive church workers in Brazil, the Indians and other effected communities, the Scarboro missionaries have been pressuring the Brazilian authorities to address the severe, dam-related pollution of the Uatumã river and its destruction of the health, and food and water supplies of the downstream populations.

Resources: The quarterly newsletter *Witnesses of Hope;* the booklet *Balbina: Catastrophe and Destruction in the Amazon.*

Contact: Daniel M. Gennarelli
Scarboro Foreign Mission Society
2685 Kingston Rd.
Scarborough, ON
M1M 1M4 Canada
Phone: (416) 261-7135

▶ **Taskforce on the Churches and Corporate Responsibility**

Objectives: The Taskforce is an ecumenical coalition that gives advice on corporate responsibility to churches, corporate shareholders, and Canadian companies.

Programs: The Taskforce is also researching the activities of Brascan mining company and other Canadian companies causing social or environmental damage in the Amazon.

Resources: Annual report

Contact: Marjorie Ross
Associate Coordinator
129 St. & Clair Ave. W.
Toronto, ON M4V 1N5
Canada
Phone: (416) 923-1758

▶ **Resource Groups**

▶ **Montreal Native Communications (MNC)**

MNC is an information center on Indian issues which caters to French speaking peoples. MNC monitors native areas throughout the Americas including the Amazon and sponsored the Native American film festival in Montreal.

Special Needs: Computer, financial support for publications.

Resources: The bimonthly newspaper *Sans Reserve* (in French)

Contact: Felix Atencio-Gonzáles
MNC
3575 Blvd. St. Laurent
Suite 513
Montreal, PQ H2L 3G4
Canada
Phone: (514) 843-6098
Fax: (514) 843-5681

NORTH AMERICA
UNITED STATES

Amnesty International USA (AIUSA)

Objectives: Amnesty's world-wide membership of one million works for the release of prisoners of conscience, campaigns for the abolition of torture and the death penalty, and opposes extra-judicial execution by governments. Under the direction of AI headquarters in England, AIUSA keeps Americans informed about prisoners and human rights violations worldwide, and directs letter-writing, media, and lobbying campaigns.

Programs: See AI (UK), page 78.

Contact: Sheila Dauer
Co-group Coordinator
Amnesty International USA
322 Eighth Avenue
New York, NY 10001
U.S.A.
Phone: (212) 807-8400
Fax: (212) 956-1157

Bank Information Center (BIC)

Objectives: BIC was started in 1987 to provide Third World public-interest groups with information about projects funded by multilateral development banks (MDBs).

Programs: BIC has been providing Amazonian groups with information on World Bank and InterAmerican Development Bank operations, projects, and policies. And as part of the MDB-reform campaign, BIC also has played a leading role in lobbying the World Bank for an environmental impact assessment process; in organizing the international forum for public-interest groups at the World Bank annual meeting in 1989; and in coordinating activities among MDB campaign groups around the world.

Resources: *Banking on Ecological and Social Disaster*, a booklet on MDB and IMF lending policies and their impacts around the world, including Brazil and Bolivia.

Contact: Chad Dobson, Director
Bank Information Center
2000 P St., NW
Suite 215
Washington, DC 20036
U.S.A.
Phone: (202) 822-6630
Fax: (202) 822-6644
Econet: bicusa

Brazil Network

Objectives: The Network is a new organization that helps build cooperation between Brazilian and U.S. popular movements.

Programs: The Network is now developing an information clearinghouse on Brazilian political, social, and environmental issues.

Resources: *Contato* newsletter (8 issues/yr.).

Contact: Linda Rabben
Brazil Network
P.O. Box 2738
Washington, DC 20013
U.S.A.
Phone: (202) 234-9382
Phone: (202) 387-7915

Catalyst

Objectives: Catalyst takes a bio-regional approach to environmental preservation work.

Programs: Catalyst educates New Englanders on rainforest preservation issues; maintains a library on corporate activities that are environmentally destructive; and is preparing an action guide, *Listening to the Forest*, that will describe threats to forests around the world, including the Amazon.

Resources: The monthly newsletter *CATALYST*.

Contact: Susan Meeker-Lowry
Director
Catalyst
64 Main St., 2nd Floor
Montpelier, VT 05602
U.S.A.
Phone: (802) 223-7943
Fax: (802) 765-4262

Conservation International (CI)

Objectives: CI is a three-year-old 30,000-member organization that works with Third World environmental groups and government agencies to preserve biodiversity and promote sustainable development.

Programs: CI negotiated the world's first debt-for-nature swap which took place in Bolivia's Beni region in 1987. The swap enables CI to work closely with the Bolivian Forest Service, the Bolivian National Academy of Science, and other groups on the management plan for the 11-million-hectare Chimane Ecosystem, which includes the 540,000-hectare Beni Biosphere Reserve. CI has been contracted by the U.S. Agency for International Development (AID) to prepare AID's management plan of Bolivia's natural resources.

In 1990, CI launched its "Rain Forest Imperative," a resource-management program to save ten "hotspots" in highly threatened tropical rainforests of "megadiversity countries." Two of the hotspots are the Bolivian Chimane Ecosystem and Colombian Amazon. The program will include collecting ethnobotanical information.

Resources: The quarterly magazine *Orion*; the monthly newsletter *Tropicus*.

Contact: Lisa Semolare
Conservation International
1015 18th St. NW
Suite 1000
Washington, DC 20036
U.S.A.
Phone: (202) 429-5660
Fax: (202) 887-5188

A Chimane Indian woman.
Photo: Courtesy of Conservation International

Above: Workers in Rio Branco load brazil nuts onto a barge for export. Photo: Kit Miller. Below: A package of Rainforest Crunch candy.

▶ **Cultural Survival (CS)**

Objectives: CS is a 16,000-member organization started in 1972 by social scientists to ensure the survival of indigenous peoples and ethnic minorities by helping them overcome threats to their lands, human rights, and livelihood. CS's projects are designed to empower native peoples with the economic, political, and technical capabilities needed for them to adapt in their own terms to the changing times. The emphasis of many CS programs is on building communication and cooperation between Indian groups through support of national and international federations.

Programs: As part of its tropical forest project, CS established the Rainforest Marketing Network to create international markets for Brazil nuts, cashew nuts, and scores of other rainforest nuts and fruits. The new demand of these goods, says CS, will encourage sustainable use of the forest, increase economic opportunities for forest dwellers, and support the growth of indigenous peoples' cooperatives.

In the U.S. and abroad, CS uses the promotion and sale of goods to educate consumers about the issue and to generate funds for CS projects. Ben & Jerry's Ice Cream (US) and the natural cosmetics company, the Body Shop (UK), were among the first businesses to place rainforest orders. More than 30 other companies are currently following their example. A successful international marketing program will enable subsistence forest collectors to earn between four and twenty times what they now make, according to CS Program Director Jason Clay. "Trade is much better than a handout," says Clay, "and it will be far more effective at protecting forest people."

Other CS Amazonian projects in Brazil include support for training the Union of Indigenous Nations (UNI) in law and natural resource management; legal assistance to the Yanomami land rights case and to the Nucleus for Indigenous Rights; CEDI's Indian lands monitoring program; and the Institute for Ethnobotany in Belem to identify products with international marketing potential. In Colombia, projects aid the native-run Puerto Rastrojo resource-management project on the Caquetá River.

CS's "Forest Residents as Forest Managers Program" works with local Indian organizations to develop native run, resource-management programs such as Pumaren in the Ecuadorian Amazon. With assistance from CS and FOIN (Federation of Indigenous Organizations of Napo), Pumaren formed a Quichua technical team to work with local Indian communities and with the government to secure land titles, and to obtain training in alternative methods of resource management (see FOIN page 55). CS supports similar projects in Colombia (see Puerto Rastrojo page 52), Bolivia,

Peru, and other parts of Latin America.

The Forest Residents Program has also facilitated information exchanges among Indians working on similar projects in Latin America as part of an ongoing effort to build an international Indian network of sustainable development experts.

Resources: *Cultural Survival Quarterly;* case studies of CS projects; special reports; 350 publications distributed for 15 other organizations; Rainforest Crunch candy (a Brazil/cashew nut brittle) and other forest products; T shirts, calendars, postcards and posters.

Contact: Ted Macdonald
Project Director
Cultural Survival
11 Divinity Ave.
Cambridge, MA 02138
U.S.A.
Phone: (617) 495-2562
Fax: (617) 495-1396

Environmental Defense Fund (EDF)

Objectives: EDF is a 100,000-member organization, whose staff of lawyers, scientists, and economists has been conducting one of the most rigorous lobbying campaigns in the U.S. for environmental protection and sustainable development. EDF's Brazil work has been focused on reforming multilateral development bank (MDB) policies, which have been responsible for financing many of the large-scale destructive development projects in Brazil.

Programs: EDF has been lobbying MDBs to hold back loans to projects such as Polonoroeste resettlement scheme, Carajas mining operations, the Acre Road construction, and the Energy Sector until Brazil takes measures to safeguard the environment and local communities. The MDBs' reluctance to do so has forced EDF to get support from the U.S. Congress and Treasury Department (which have oversight over US-MDB contributions) by lobbying, testifying before Congress, and launching U.S. letter-writing campaigns involving millions of Americans from all major U.S. environmental groups. EDF also sponsors visits of Brazilian activists—such as Kayapó chiefs and rubber tappers—to Washington to lobby MDBs and Congress themselves.

Fighting destructive projects is only one part of the EDF's MDB campaign; the other is finding sustainable development alternatives to replace them. "In this game," says EDF's International Director Bruce Rich, "doing your homework thoroughly is a prerequisite for effective action." This is why EDF is now working with a University of São Paulo team to develop an alternative energy plan for Brazil that relies on conservation and end-use efficiency rather than on 130 hydroelectric dams planned by the government.

EDF has also taken a leading role in the United States to work for the establishment of extractive reserves. EDF helped Chico Mendes lobby the MDBs and the U.S. Congress during his 1985 visit to Washington—an effort that boosted international support for the rubber tappers and contributed to the subsequent establishment of more than 15 extractive reserves. After Mendes's death, EDF created the Chico Mendes Fund (see page 75), which raises money for the work of the National Council of Rubber Tappers.

Bruce Rich of EDF testifying before the U.S. Congress on Amazonian deforestation.
Photo: ©Aguirre/Switkes

Resources: The bimonthly newsletter *EDF Letter;* dossiers on MDB projects in Brazil; reports on extractive reserves and the rubber tappers.

Contact: Bruce Rich
International Director
or Steve Schwartzman
Staff Anthropologist
EDF
1616 P St., NW
Washington, DC 20010
U.S.A.
Phone: (202) 387-3500
Fax: (202) 234-6049
Telex: 6503232147

Environmental Policy Institute/Friends of the Earth/Oceanic Society (EPI/FOE/OS)

Objectives: In 1989, the EPI merged with FOE and OS to create one 45,000-member organization. As one of the organizations involved in the multilateral development bank (MDB) campaign, EPI/FOE/OS lobbies MDBs and the U.S. Congress to reform MDB lending policies and support sustainable development.

Programs: EPI/FOE/OS distributes information on major MDB projects in the Amazon and other regions to organizations around the world via computer networks. EPI/FOE/OS has also taken a leading role in investigating the environmental impacts of the International Monetary Fund's lending policies and coordinating an international campaign to reform them. The organization is also working to reform the Tropical Forest Action Plan, a blueprint adopted by the governments of more than 50 countries—including Amazon nations—for managing the tropical forests. EPI has been pushing for more local organizations and indigenous peoples' involvement in the multimillion-dollar Action Plan's implementation.

EPI/FOE/OS also runs a small grants program (up to $5,000) for Third World organizations who are fighting to stop tropical deforestation.

Resources: The bimonthly newspaper *Not Man Apart;* campaign packets; *International NGO Directory* of people active in tropical forest preservation and the MDB campaign.

Contact: Alex Hittle
International Coordinator
EPI/FOE/OS
218 D St., SE
Washington, DC 20003
U.S.A.
Phone: (202) 544-2600
Fax: (202) 543-4710
Telex: 62949875

Global Exchange

Objectives: Global Exchange builds cooperation between grassroots groups in the U.S. and developing countries.

Programs: The Exchange sponsors an annual "reality tour," which takes Americans to Brazil to make contact with popular movements. It also offers Third World groups fundraising, technical, and research assistance on environmental, political, and development issues.

Contact: Medea Benjamin
Director
Global Exchange
2141 Mission Street
Room 202
San Francisco, CA 94110
U.S.A.
Phone: (415) 255-7296

Greenpeace USA

Objectives: Greenpeace is a direct action and advocacy environmental organization with offices in 25 countries and more than three million supporters worldwide. Greenpeace is best known for protecting marine mammals, opposing nuclear-bomb testing, and fighting toxic pollution. In early 1990, Greenpeace launched a new effort to combat destruction of tropical rainforests.

Programs: The Amazon will be an initial focus for the Greenpeace Tropical Forest Campaign (TFC), when the new Greenpeace Brazil office opens in 1990. Staffed by Brazilians, Greenpeace Brazil will work with other Brazilian organizations to develop strategies for tackling key environmental problems, such as waste trade, and toxic dumping, and deforestation (commercial logging, cattle ranching, and inappropriate agriculture). TFC will also promote extractive and indigenous reserves and rehabilitation of damaged forest areas. It will have campaigners in Europe, the U.S., and Japan who will investigate and oppose bank, aid agency, and private-interest projects that destroy or degrade the Amazonian forests.

Resources: *Greenpeace Magazine*, bimonthly.

Contact: Campbell Plowden
Coordinator
Tropical Rainforest
Campaign/Greenpeace
1436 U St. NW
Washington, DC 20009
Phone: (202) 462-1177
Fax: (202) 462-4507
Telex: 89-2359

Indian Law Resource Center (ILRC)

Objectives: Established and directed by Native Americans, the 12-year old ILRC provides free legal assistance to Indian nations and tribes in their efforts to combat racism in the law and the loss of their land and human rights. The center also helps present cases of human rights violations to international bodies such as the UN.

Programs: ILRC has provided Amazonian Indian and grassroots organizations with handbooks on Indian rights; worked closely with the Coordinating Body of the Indigenous Organizations of the Amazon Basin (COICA) in the UN Working Group on Indigenous Populations; and is now helping the Ecuadorian Indian confederation (CONFENIAE) and Brazil's Union of Indigenous Nations (UNI) to publicize the human rights violations in the Amazon.

Resources: *Handbook on International Human Rights Complaint Procedures;* and an annual report.

Contact: Tim Coulter
Executive Director
ILRC
601 E St., SE
Washington, DC 20003
U.S.A.
Phone: (202) 547-2800

IRN President Phil Williams holding *World Rivers Review* while protesting the financing of the Altamira dams at Citicorp Bank in San Francisco, California. Photo: A. Gennino

International Rivers Network (IRN)

Objectives: IRN formed in 1985 to stop the mismanagement and destruction of the world's watersheds and river systems caused by industrial development, pollution, deforestation, and channelization. IRN's focus has been on halting the construction of large-scale projects, such as high dams, which have devastating social and environmental impacts. IRN is a clearinghouse for information on the world's river development projects and provides technical assistance and grassroots support to the local groups fighting them.

"Rivers are the planet's lifeblood," says hydrologist Philip Williams, cofounder and president of IRN. "Their destruction is as important an element of the global environmental crisis as the destruction of forests, soils, atmosphere, and marine ecosystems. In the Third World such destruction, caused especially by construction of large dams, can be the driving force in the vicious cycle of inappropriate development, environmental degradation, impoverishment, and economic collapse. IRN is here to bring an end to this destruction and to promote the sustainable management of our fresh-water resources."

Programs: IRN's Amazon focus has been on fighting Brazil's Plan 2010, which calls for the construction of at least 70 high dams in the Amazon basin. In 1988, at IRN's international conference on river issues, Latin American delegates formed a regional network to keep each other better informed about the dams and other water projects in their areas. This network is now being coordinated by the Union to Protect the National Environment (UPAN).

In 1989, IRN and the Rainforest Action Network co-sponsored demonstrations against the World Bank for financing the Plan 2010. At the same time, IRN launched a campaign to step up public pressure against dam construction in the Amazon, while promoting a more comprehensive and sustainable approach to water-resource management.

Resources: The bimonthly newsletter, *World Rivers Review*; the two-volume book, *Environmental and Social Effects of Large Dams*, by Edward Goldsmith and Nicholas Hildyard.

Contact: IRN
301 Broadway, Suite B
San Francisco, CA 94133
U.S.A.
Phone: (415) 986-4694
Fax: (415) 398-2732
Telex: 6503532706
Econet: irn

IRN staff get through another day of saving the world's rivers. Standing: Amy Chueng and Owen Lammers. Middle: Juliette Majot. Photo: A. Gennino

▶ National Wildlife Federation (NWF)

Objectives: With its 5 million members, NWF is America's largest wildlife conservation organization. NWF's focus has been on preserving the quality of America's environment, particularly its wilderness areas. But in the last few years, NWF has devoted more attention to global environmental problems, particularly those caused by U.S. funding of ecologically destructive development projects in Third World countries.

Programs: As part of the Multilateral Development Bank (MDB) campaign, NWF informs its members of the devastating impacts on Brazilian Amazon from MDB projects (Polonoroeste resettlement scheme, Carajás mines, and Energy Sector's high dams). NWF conducts lobbying and letter-writing campaigns directed at MDBs and the U.S. Congress to stop project funding until environmental guidelines for those projects are established and followed by Brazil. NWF's international director, Barbara Bramble, was instrumental in formulating the debt-for-nature swap strategy.

Resources: The bimonthly magazines *Wildlife Magazine* and *International Wildlife*.

Contact: Barbara Bramble
International Director
NWF
1400 16th St NW
Washington, DC 20036
U.S.A.
Phone: (202) 797-6600
Fax: (202) 797-6646
Telex: 6502657521

▶ The Nature Conservancy

Objectives: With 550,000 members, the Conservancy is one of the largest U.S. environmental organizations and responsible, it says, for saving 4,000 hectares of the natural world each day. In the U.S., the 40-year-old organization is best known for protecting sensitive wildlife habitats by purchasing them outright. The group's motto: "No piece of land is too small or too large to be protected." Today, the Conservancy owns 2.5 million hectares, mostly in the U.S., and manages projects totalling 13.5 million hectares in the Americas and Caribbean.

Programs: In Latin America, the Conservancy usually does not purchase land but often works with the U.S. AID and other development agencies to provide local organizations with the finances and training needed to manage endangered wildlands. In 1989, funds raised with World Wildlife for the 1989 debt-for-nature swap were earmarked for park management in Ecuador and Peru.

The Conservancy's Amazonian projects come under "Parks in Peril," a campaign to save 200 key tropical areas covering more than 400 million hectares throughout Latin America and the Caribbean. The U.S. Congress is granting $2 million for the first 20 sites—a sum that will be matched by private dollars. The money will be made available to various environmental organizations, such as BIOMA (Venezuela), which is planning the management of La Neblina National Park.

Resources: Conservation Data Center computerized information service on biological diversity and other environmental issues; *Diversidata*, the Conservation Data Center's monthly bulletin (in Spanish); the bimonthly *Nature Conservancy Magazine*.

Contact: Latin America Program
The Nature Conservancy
1815 North Lynn St.
Arlington, VA 22209
U.S.A.
Phone: (703) 841-5300
Fax: (703) 841-1283

Left: Paiakan Kayapó spoke out against the Altamira dams on his 1988 U.S. tour.
Photo: ©Aguirre/Switkes.
Top: A tapir rambling through the rainforest.
Photo: courtesy of Conservation International.

▶ New England Tropical Forest Project

Objectives: The Project is a 1,500-member organization that builds awareness of and support for rainforest protection among New Englanders.

Programs: The Project conducts letter-writing campaigns to pressure governments, corporations, and banks to stop funding destructive projects. In September 1989, The Project sponsored the "Arctic to Amazonia" conference that brought native peoples of the Americas and the U.S. public together to discuss deforestation problems. The next conference (1990) will look at indigenous examples of sustainable-resource management.

Resources: the quarterly newsletter *Tapper*.

Contact: Erik van Lennep
Director
New England Tropical
Forest Project
P.O. Box 73
Strafford, VT 05072
U.S.A.
Phone: (802) 765-4337
Fax: (802) 765-4262

▶ Oxfam America

Objectives: Like the six other autonomous Oxfam groups around the world, Oxfam America has become one of the leading support groups for grassroots development projects in the Third World. In the U.S., Oxfam educates Americans on development and hunger and raises funds for specific projects abroad.

Programs: In fiscal year 1989, Oxfam America raised nearly $2 million for Latin American projects, many of which were located in the Amazon: the Indian Health Program Exchange Visit in Acre, Brazil; the Amazonian Indian confederation's (COICA) international work; the evaluation of Indian land protection plans; organizational development of Indigenous Confederation of the Bolivian Amazon (CIDOB); technical training for the Inter-Ethnic Development Association of the Peruvian Jungle (AIDESEP); network development of Amazonian Indians in Venezuela; and the Five Centuries of Resistance campaign in Ecuador.

In 1989, Oxfam funding enabled a COICA delegation to tour the U.S. and discuss deforestation issues with American environmentalists and urge them to work more closely with native peoples of the Amazon on future campaigns.

Resources: *Oxfam America News* (3 issues/yr.).

Contact: Juan Aulestia
South America Programs
Representative
Oxfam America
115 Broadway
Boston, MA 02116
U.S.A.
Phone: (617) 482-1211
Fax: (617) 338-0187
Telex: 94-0288

▶ Project Abraço

Objectives: Abraço supports grassroots movements in Brazil, mainly through information campaigns in the United States on the misuse of aid, the crippling effects of foreign debt, and environmental disasters.

Programs: Abraço also cosponsors "reality tours" with Global Exchange, which take Americans to Brazil for direct contact with grassroots movements.

Resources: *Terra Nossa* newsletter; *Brazil's Debt and Deforestation–A Global Warning*, a six-page action alert produced by Judith Hurley and the Institute for Food and Development Policy; information packets and maps; workshops and audio/visual programs.

Contact: Judith Hurley
Coordinator
Project Abraço
515 Broadway
Santa Cruz, CA 95060
U.S.A.
Phone: (408) 425-5939

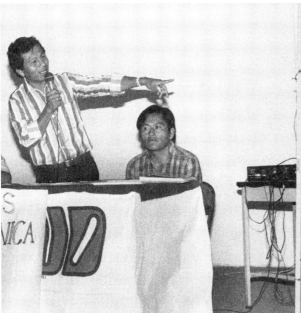

Oxfam has supported the international work of COICA.
Above: COICA delegates meet in Bolivia for a strategy session.
Top right: Cristobal Tapuy of CONAIE (Ecuador).
Middle: Venezuelan delegate and José Luis Xeretê of Brazil.
Bottom: Gustavo López of ONIC (Colombia). Photos: ©Aguirre/Switkes

▶ **Rainforest Action Network (RAN)**

Objectives: RAN is a 30,000 member grassroots organization dedicated solely to stopping the rapid destruction of the world's tropical rainforests. RAN works with local communities and organizations in more than 60 countries to pressure corporations, governments, aid agencies, and banks to stop funding destructive projects. RAN relies on hard-hitting media campaigns, demonstrations, and letter-writing campaigns to respond quickly and directly to the forces that threaten the rainforests.

"We're not afraid of direct confrontation with the biggest companies, the banks, or the U.S. government if it means saving rainforests and protecting the human rights of the people who live in them," says RAN Director Randall Hayes. "Rainforests are disappearing at a frightening rate and we simply can't let it continue."

Programs: RAN's 150 rainforest action groups (RAGs) located throughout the U.S. get local press coverage of and community support for rainforest issues. RAN regularly attracts national press coverage in the U.S. by organizing demonstrations and consumer boycotts directed against companies destroying rainforests; sponsoring benefit rock concerts; and accompanying U.S. congressional committees to rainforest areas. RAN's other effective vehicle for national coverage is placing full-page ads in top U.S. newspapers such as *The New York Times* and *The Wall Street Journal*, describing the plight of the rainforest and urging readers to take action.

On the Amazon front last year, RAN sponsored benefits for Kayapó Indian leaders from Brazil who are fighting high-dam projects in their homeland. In September 1989, RAN co-sponsored two demonstrations at the World Bank's annual meeting in Washington to protest Bank funding of hydro dams in the Amazon. One of the demonstrations included a blockade of the building where the meeting was taking place, which resulted in the arrest of 13 protesters.

In 1990, RAN's Amazon projects include fundraising for grassroots Brazilian organizations; the co-sponsoring of an "eco-tour" of the Amazon with the river rafting company Sobek; and the investigation of U.S. oil drilling operations in the Amazon, which may result in a campaign to stop Conoco Corp.'s activities in Ecuador. And as part of its Tropical Timber Campaign, RAN is calling for a U.S. ban on the import and use of all tropical hardwood products except those that come from a forest reserve where:
▲ Local communities participate in all stages of planning and implementation of extraction;
▲ Timber projects do not interfere with land rights of the local communities; and
▲ New areas of timber production are created through reforestation and regeneration of severly damaged areas.

RAN staff clockwise from bottom left: Maria Ines Salgado, Amy Cheung, Ian McWilliam, Annie Szvetecz, Pam Wellner, Darin Plutchok, Suzanne Head, Bruce Chamberlain, Oona Smith, Katie Montgomery (in back), Randall Hayes, and Francesca Vietor.
Photo: A. Gennino

Amazonia: Voices from the Rainforest staff from front: Kara Adanalian, Angela Gennino, Beto Borges, Monti Aguirre, Steven Barton. Back: Glenn Switkes, Scott Desposato. At large in San Francisco: Anneke Vonk, Belinda van Valkenburg, Bill Walker; in Brazil: Maria Souza, Brent Millikan.
Photo: Pam Wellner

Resources: Quarterly newsletter *World Rainforest Report;* monthly *Action Alerts;* information packets, slide shows and films; T-shirts.

Contact: Suzanne Head
Program Manager
RAN
301 Broadway, Suite A
San Francisco, CA 94133
U.S.A.
Phone: (415) 398-4404
Fax: (415) 398-2732
Telex: 15127-6475
Econet: RAINFOREST

Rainforest Alliance

Objectives: The Alliance promotes rainforest conservation by stressing the medicinal uses of the forest plants and developing methods for sustainable logging of tropical timber.

Projects: The Alliance is the North American representative of AMETRA 2001, a health project in the Peruvian Amazon that uses of traditional and western medicine to improve health care among the Indians. The Periwinkle project is another Alliance project recently launched to educate health professionals on the importance of the rainforest's plants in healing. The Alliance's Tropical Timber Project is based on work with the tropical timber industry to develop sustainable logging practices. The Alliance held an international forum in April 1989 on tropical timber issues which resulted in a long-range program to develop a set of industrial guidelines for sustainable forestry and using them to evaluate timber concessions in Southeast Asia and the Amazon, which supply U.S. markets. The Alliance also offers a $30,000 two-year fellowship to researchers working in agroforestry.

Resources: The quarterly newsletter *Canopy*; an events calendar; a library; and an educational curriculum.

Contact: Daniel R. Katz
Rainforest Alliance
270 Lafayette St., Suite 512
New York, NY 10012
U.S.A.
Phone: (212) 941-1900,
Fax: (212) 941-4986

Rainforest Foundation

Objectives: The rock musician Sting formed the Rainforest Foundation to help Amazonian Indians obtain legal title of their territories so that they can manage and protect their own lands.

Projects: With the help of more than 20 sister groups around the world, RF educates the public and raises funds to campaign in Brazil for the designation of a Kayapó reserve and to finance resource management, health, and education programs as part of a Kayapó sustainable development project. In 1989, RF also funded an emergency air lift of medical aid to the Yanomami, a tribe in northern Brazil whose health is rapidly deteriorating from contact with gold panners and government neglect. The airlift was conducted by the Kayapó and the Union of Indigenous Nations (UNI).

Resources: Press packet, educational programs.

Contact: Mary Daly
Executive Director
Rainforest Foundation
1776 Broadway, 14th Floor
New York, NY 10019
U.S.A.
Phone: (212) 767-1630
Fax: (212) 582-6513

Rainforest Futures

Objectives: Rainforest Future raises awareness in Santa Cruz, California about tropical deforestation and mobilizes people to support sustainable development alternatives.

Projects: RF supports the Rainforest Action Network's tropical timber campaign by urging local stores to stop selling tropical hardwoods and by lobbying for a county ban on the products. In 1990, RF plans to fund a sustainable development project in the Brazilian Amazon with proceeds from art sales.

Resources: *Rainforest Hotline*, monthly newsletter on sustainable projects in the Amazon and how Americans can support them.

Contact: Margie Manners
President
518 Meder St.
Santa Cruz, CA 95060
U.S.A.
Phone: (408) 426-9251

Sierra Club

Objectives: The Sierra Club is a 550,000-member conservation organization devoted to preserving the quality of America's environment, particularly its wilderness areas, and to solving global environmental problems caused by U.S. funding of ecologically destructive development projects in Third World countries. The Club is a leading member of the international campaign to reform the environmental policies of the multilateral development banks (MDBs).

Projects: The Club conducts lobbying and letter-writing campaigns directed at MDBs and the U.S. Congress to stop funding destructive projects abroad, including the Brazilian Amazon until governments follow environmental guidelines. In 1989, the Club backed new U.S. legislation directing MDBs to submit environmental impact assessments (EISs) on their projects to the U.S. Treasury for review prior to loan approval. The legislation will go into effect in 1991, giving the public access to the EISs and the ability to influence the U.S. vote on the project loan.

In the meantime, the Club is pressuring the U.S. Congress to have the U.S. National Environmental Protection Act (NEPA) apply to all extra-territorial activities of the U.S. government, and to limit Congressional and World Bank support for the Tropical Forest Action Plan, which promotes commercial forestry development in the world's rainforests.

Resources: The monthly *Sierra* magazine, the bimonthly *National Newsletter*; monthly *Hot Spot* action alerts; biodiversity factsheet; Conservation Campaign Brief on the MDB campaign; the video and slide show *The Tropical Rainforest: Diverse, Delegate, and Disappearing*. All are available from the Sierra Club's Public Affairs Office; 730 Polk St., San Francisco, CA 94108, U.S.A. Phone: (415) 776-2211.

Contact: Larry Williams
Washington Director
International Program
Sierra Club
408 C St. NE
Washington, DC 20002
U.S.A.
Phone: (202) 547-1144
Fax: (202) 547-6009
Telex: 4900005633
Econet: SCDC1

Sierra Club Legal Defense Fund (SCLDF)

Objectives: SCLDF is an independent, nonprofit environmental litigation group that for 19 years has been working primarily on U.S.-based environmental lawsuits. In 1989, SCLDF took on its first international case, which involved the destruction of Indian lands by U.S. oil drilling operations in Ecuador.

Projects: SCLDF is working with the Ecuadorian Indian federation, CONFENAIE, on a petition to the Organization of American States to support the Huaorani's human right to keep their traditional lands free of development. SCLDF is also bringing the Huaorani case to the UN Subcommission on the Prevention of Discrimination and Protection of Minorities. In 1990, SCLDF will be investigating litigation and other legal actions against Conoco Corp. and the Ecuadorian government for violating international human rights laws in their treatment of the Huaorani.

Resources: *In Brief* newsletter.

Contact: Vawter Parker
Coordinating Attorney
SCLDF
2044 Fillmore St.
San Francisco, CA 94115
U.S.A.
Phone: (415) 567-6100
Fax: (415) 567-7740

▶ South and Meso American Indian Information Center (SAIIC)

Objectives: SAIIC builds communication links between Indian organizations throughout the Americas and educates Americans about Indians' struggles to protect their lands, culture, and their right to self-determination.

Programs: SAIIC helps U.S. public-interest groups make contacts within the Indian communities. It has organized speaking tours throughout North America for visiting Amazonian Indian delegations, such as the one by the Coordinating Body of the Indigenous Organizations of the Amazon Basin (COICA) in October 1989. Currently, SAIIC is working with the Union of Indigenous Nations (Brazil) to send a delegation of North American Indian spiritual leaders to the Amazon. With Indian confederations of Ecuador (CONAIE) and Colombia (ONIC), SAIIC also is organizing a media campaign and North American Indian participation in the Five Hundred Years of Resistance, a series of events in 1992 to present the native interpretation of Columbus's "discovery" of America.

Resources: The quarterly *SAIIC Newsletter* (in English and Spanish); Urgent Action Bulletins; reports by Indian groups about human rights violations.

Contact: Nilo Cayuqueo
Coordinator
SAIIC
P.O. Box 7550
Oakland, CA 94707
U.S.A.
Phone: (415) 834-4263
Fax: (415) 834-4264
Telex: 154205417
Peacenet: cdp:SAIIC

▶ Survival International USA (SIUSA)

Objectives: Founded in 1969 in England, Survival is a leading international organization devoted to supporting tribal peoples' rights to survival, self determination, traditonal lands, and access to adequate natural resources. As Survival's American chapter, SIUSA's main role is to establish local contact with Survival members in the U.S. and to expand advocacy work through membership development and public education.

Programs: See Survival's International Secretariat (England) page 80.

Resources: The bulletin *Notes from SIUSA*.

Contact: Mary George Hardman
Executive Director
SIUSA
2121 Decatur Place, NW
Washington, DC 20008
U.S.A.
Phone: (202) 265-1077
Fax: (202) 265-1297

A sloth out and about.
Photo: ©Aguirre/Switkes

▶ Wildlife Conservation International (WCI)

Objectives: WCI has been among the most active international conservation organizations in the U.S. for the last 100 years. WCI now supports 90 field projects in 45 countries, including Amazonian nations, that are managed and staffed predominantly by local experts.

Programs: Within the Amazon of Brazil, Peru, Ecuador and Venezuela, WCI is organizing training workshops and field courses for local professionals to manage and protect ecosystems. WCI is also conducting research projects in national parks and extractive reserves and supporting environmental education.

Resources: The quarterly *WCI Newsletter*.

Contact: Dr. Mary Pearl
Assistant Director
WCI
NY Zoological Society
The Bronx Zoo
New York, NY 10460
U.S.A.
Phone: (212) 220-5155
Fax: (212) 220-7114
Telex: 428279 NYZWCI

▶ World Wildlife Fund & Conservation Foundation

Objectives: WWF is the largest international wildlife organization with an annual budget of more than $29 million, 737,000 members, and a network of 25 organizations around the world. WWF and its policy group support scientific research, monitors international wildlife trade, promotes sustainable development, funds local wildlife groups, and works in partnership with governments, aid agencies, the timber industry, and multilateral development banks to formulate environmental policies and projects.

Programs: In the Brazilian Amazon, WWF provides funds for the Goeldi Museum; resource-management training for extractive products; and research on freshwater turtles, forest fragments, selective logging, and primates.

In the Andean Amazon, WWF provided support for Colombia's Cahuinari National Park; primate ecology and ethnobotany in Ecuador's Cuyabeno Wildlife Reserve; a debt-for-nature swap that has funded management programs for Ecuador's Sangay, Cayambe-Coca, Podocarpus, Yasuni, and Cuyabeno national parks; and support for Peru's Manu National Park, Pacaya Samiria and Rio Abiseo reserves. WWF is also providing financial and technical support to the Peruvian Yanesha Forestry Cooperative—one of the most innovative experimental projects in native-run agroforestry (see Yanesha Cooperative page 59).

Resources: The bimonthly newsletters, *Focus* and *WWF Letter*; the books *Whose Business Is It?* (on the wildlife trade), *Power to Spare: The World Bank and Electricity Conservation*, and *Government Policies and Deforestation in Brazil's Amazon Region*.

Contact: Latin America Division
WWF & CF
1250 24th Street NW
Washington, DC 20037
U.S.A.
Phone: (202) 293-4800,
Fax: (202) 293-9211
Telex: 64505 PANDA

Yanomamo Survival Fund (YSF)

Objectives: YSF was founded in 1989 by anthropologist Napoleon Chagnon to support grassroots development projects of the Yanomamo Indians in Venezuela.

Programs: YSF works to provide financial and technical assistance in health, agriculture, and marketing areas to Venezuela's Yanomami communities' development projects, particularly SUYAO handicrafts cooperatives (see SUYAO page 63).

Contact: Napoleon Chagnon
President
YSF
405 Calle Granada
Santa Barbara, CA 93105
U.S.A.
Phone: (805) 682-8497

Resource Groups

Amazon Research and Training Program (ARTP)

ARTP offers a graduate-level seminar/field-research program in on Amazon environmental topics in cooperation with the Federal University of Acre in Brazil. ARTP wants to receive new publications and help Amazonian groups meet their research needs.

Resources: The biannual *ARTP Newsletter*, which includes a list of researchers and new publications; an up-to-date data base on Amazon tropical conservation and development issues.

Contact: Dr. Marianne Schmink
Executive Director
ARTP
Center for Latin American Studies
Grinter Hall
University of Florida
Gainesville, FL 32611
U.S.A.
Phone: (904) 392-0375
Fax: (904) 392-9506

Ashoka: Innovators for the Public

Ashoka funds individuals who are public-interest entrepreneurs. Amazonian fellowships have been awarded to Alba Lucy Figueroa, a health educator for tribal groups; Mary Allegretti, director of the Institute for Amazonian Studies, who is developing a legal framework for extractive reserves; Indian leader Ailton Krenak to improve the living conditions and legal rights of Brazil's indigenous peoples; and Chico Mendes for economic development programs for rubber tappers.

Contact: Miriam Parel
Director
Ashoka
1200 North Nash St.
Arlington, VA 22209
U.S.A.
Phone: (202) 628-0370
Fax: (202) 628-0376

Better World Society (BWS)

BWS produces TV programs on the environment, development, and the arms race. BWS has recently developed a focus on the Amazon by producing a documentary on Chico Mendes; funding the National Council of Rubber Tappers in Brazil; and working with the Brazilian media. BWS maintains a TV library on environmental programs and is looking for coproduction partners.

Resources: The quarterly *Better World Letter*.

Contact: Thomas Belford
Executive Director
Better World Society
1100 17th St., NW
Suite 502
Washington, DC 20036
U.S.A.
Phone: (202) 331-3770
Fax: (202) 331-3779

Center for Indigenous Knowledge for Agriculture and Rural Development (CIKARD)

CIKARD collects information on indigenous knowledge and uses of the environment and applies it toward new strategies for sustainable development.

Resources: The quarterly bulletin *CIKARD News*; an information center and library.

Contact: Dr. D. Michael Warren
Director
CIKARD
324 Curtiss Hall
Iowa State University
Ames, IA 50011
U.S.A.
Phone: (515) 294-0938
Fax: (515) 294-0907
Telex: 283359 IASU UR

Chico Mendes Fund (CMF)

After the murder of Chico Mendes, the Environmental Defense Fund (EDF) formed the CMF to raise money for the National Council of Rubber Tappers' campaigns for extractive reserves, community development projects, and alliances with other other peoples of the forest. Donations to CMF are tax deductible and should be made out to the Environmental Defense Fund (earmarked for the CMF) and mailed to:

Contact: Steve Schwartzman
EDF
1616 P St., NW, Suite 150
Washington, DC 20036
U.S.A.
Phone: (202) 387-3500
Fax: (202) 234-6049

Institute for International Cooperation and Development (IICD)

IICD develops the leadership skills of social, environmental, and political activists in cooperation with Third World movements, such as those of Brazil's landless peasants. Classes are conducted in the U.S and abroad.

Resources: Course catalogs and newsletter.

Contact: Julie Sweedler
Acting Director
IICD
P.O. Box 1063
Amherst, MA 01004
U.S.A.
Phone: (413) 268-9229
Fax: (413) 268-3350

International Society of Tropical Foresters (ISTF)

ISTF works for the protection and wise management of tropical rainforests by supporting innovative research and organizing workshops and symposia.

Resources: The quarterly bulletin *ISTF News* (in English and Spanish); a list of ISTF members.

Contact: Warren T. Doolittle
President
ISTF
5400 Grosvenor Lane
Bethesda, MD 20814
U.S.A.
Phone: (301) 897-8720

Medicina da Terra
(Earth Medicine)

Medicina da Terra is a new organization that educates health professionals on the important role rainforest flora and fauna play in health and healing. Medicina plans to purchase a tract of rainforest in Acre, Brazil, for studying medicinal plants and for rubber tappers and Brazil-nut gatherers to extract their products.

Contact: David Lenderts, M.D.
Medicina da Terra
P.O. Box 315
Norwood, CO 81423
U.S.A.
Phone: (303) 327-4689

Missouri Botanical Gardens

The Gardens conducts botanical and ethnobotanical research in the Amazon areas of Venezuela, Colombia, Ecuador, Bolivia, and Peru.

Resources: The bimonthly *Missouri Botanical Gardens Bulletin*.

Contact: Enrique Forero
Director of Research
Missouri Botanical Gardens
P.O. Box 299
St. Louis, MO 63166
U.S.A.
Phone: (314) 577-5111
Fax: (314) 577-9596

N.Y. Botanical Gardens (NYBG)

NYBG is an institution with a unique focus on applied botanical research. Through its Institute of Economic Botany, NYBG studies in the Amazon are geared to helping the local populations learn how to sustainably utilize tropical flora. The projects now underway include a study on native fruit trees in the Loreto region of Peru; a marketing analysis for regional forest products sold in Iquitos; the production of medicines from local plants in Brazil's eastern Amazon; and in the same region, technology development for harvesting *babassu* palm and production of an education manual on economic palms for a school on the extractive reserve.

Resources: NYBS reports include *New Directives in the Study of Plants and People; Resource Management in Amazonia: Indigenous and Folk Strategies; Advances in Economic Botany: Swidden, Fallow, and Agroforestry in the Peruvian Amazon;* and the manual *Economic Plants of the Oriente of Ecuador*.

Contact: Doug Daly
Curator of Amazonian Botany
New York Botanical Gardens
200 St. & Southern Blvd
Bronx, NY 10458
U.S.A.
Phone: (212) 220 8700
Fax: (212) 220-6504

Rainforest Health Alliance (RHA)

RHA builds support for rainforest preservation among people in the health, food, and art professions by bringing them into contact with the Indians who use the rainforest for medicinal, traditional, and economic purposes. In 1990-91, an RHA-sponsored expedition will visit biological reserves in the Amazon and attend workshops on rainforest plants at the Union of Indigenous Nations' Center for Research and Resource Management in Goiania.

Resources: Fact sheets on medicinal plants in Amazon with an emphasis on their traditional uses.

Contact: Pam Roberts
Director
RHA
Fort Mason, Bldg. E, #205
San Francisco, CA 94123
U.S.A.
Phone: (415) 921-1203

Smithsonian Institution

The Smithsonian, founded by Congress in 1984, is the world's largest educational/research complex with 84 percent of its $320 million annual budget coming from the government. In the U.S. the Smithsonian has put together an impressive traveling exhibition on the tropical rainforest, and at Smithsonian headquarters, is combining the elements of zoos, aquariums, museums, botanical gardens, and arboretums to create a huge Amazonia ecosystem exhibit at the National Zoological Park. Smithsonian sponsors extensive field research in wildlife, ethnobotany, and anthropology, and has organized tours for special interest groups such as congressional delegations.

Resources: The book *Saving the Tropical Forest*.

Contact: Judy Gradwohl
Latin American Division
Smithsonian Institution
Washington, DC 20560
U.S.A.
Phone: (703) 357-4797

Threshold Foundation

Threshold has funded several conferences in the U.S. on tropical rainforest preservation and Brazil's Union of Indigenous Nations (UNI). It wants to take a more active role in Amazonian issues such as the sponsoring of a computer information network.

Contact: George Binney or
Kathy Daly
Threshold Foundation
P.O. Box 18
Tumacaco, AZ 85640
U.S.A.
Phone: (602) 432-7353 or 398-2782

World Resources Institute (WRI)

World Resources Institute is a policy and research organization created to help governments, international organizations, and the private sector address the issues of economic growth and environmental integrity.

WRI prepared the Tropical Forest Action Plan with the World Bank, the Food and Agriculture Organization, and the UN Environment Programme. Adopted by 56 countries, the Plan lays out a strategy for tackling deforestation problems and determining timber investment priorities.

WRI has recently started "Gift to the Future," another sweeping international program with the multilateral development banks, scientific institutions, and numerous wildlife organizations to develop a world-wide strategy for protecting biodiversity.

Resources: The monthly bulletin *NGO Reporter*; *World Resources*, a book series on the state of the earth's natural resources; the book *The Forest for the Trees? Government Policies and the Misuse of Forest Resources*.

Contact: WRI
1709 New York Ave., NW
Washington, DC 20006
U.S.A.
Phone: (202) 638-6300
Fax: (202) 638-0036
Telex: 64414 WRIWASH

European Environmental Bureau/Working Group on Amazonia

Objectives: European Environmental Bureau is comprised of 100 organizations from the 12 member states of the European Community (EC). Its task is to lobby the EC on issues concerning the European environment and areas abroad that are affected by EC aid or loans.

Programs: The Bureau has a Working Group (WG) on Amazonia, which has been urging the EC to stop loans to the Carajás iron ore project in Brazil and promoting alternatives for charcoal-fueled pig iron production. Through a traveling exhibition, letters to politicians, and other efforts, the Working Group has focused public attention on the Yanomami and health and land issues and raised funds for establishing extractive reserves.

WG's goal is to support sound development in the Amazon by creating European markets for sustainably produced goods; and pressuring European governments and banks to completely restructure the debt and to fund programs in energy conservation, reforestation, and crop substitution in coca growing areas. WG also provides funds for Amazonian activists' speaking tours of Europe.

Zoró Indian of Brazil.
Photo: ©Aguirre/Switkes

Resources: A directory of private and governmental organizations and agencies involved in Amazonian activism and research (forthcoming in 1990).

Contact: Wouter Veening
Chairman

European Environmental Bureau
20, Rue du Luxembourg
Brussels
B 1040 Belgium
Phone: (02) 514-1250 or 514-1432
Fax: (02) 514-0937
Telex: BEE 62 720

EUROPE
AUSTRIA
BELGIUM

Gesellschaft für Bedrohte Volker-Osterreich (GBV)

(Association for Endangered Peoples-Austria)

Objectives: GBV-Austria is one of the three GBVs located in German-speaking countries (the others are in West Germany and Luxembourg), which support human rights for ethnic minorities and indigenous peoples around the world by conducting demonstrations, letter-writing and lobbying campaigns, and international seminars.

Programs: GBV has established an Amazon information center on corporate investments, militarization, colonization, and public-interest groups working on conservation and sustainable development. GBV also intends to start a twinning project that will link Austrian communities to Amazonian communities needing support for development projects. In the fall of 1989, GBV held an international seminar on Brazilian Amazonian issues of hydro-power development, the Carajás mining project, and Indian rights. In the fall of 1990, GBV will sponsor an Amazon exhibition in the Austrian Parliament for Minorities and Ecology Week.

Resources: The journals *Pogrom* and *Bedrohte Völker (Endangered Peoples)*.

Contact: Carlos Macedo
Campaign Director
Mariahilfer Str. 105/11/13
A-1060 Wein
Austria
Phone: (222) 597-1176
Fax: (222) 597-3743

Stuengroep voor Inheemse Volkeren (KWIA)

(Support Group for Indigenous Peoples)

Objectives: KWIA is a 300-member, two-year-old group, that supports indigenous people's struggles for self-determination, cultural survival, and land rights.

Programs: KWIA is producing a series of dossiers on Brazilian Indians and Belgium's connection to their oppression and rainforest destruction. The group plans to have a representative working in northern Brazil with a Yanomami community. At the request of an Amazonian Indian organization, KWIA will lobby Belgian government institutions and members of the Belgian and European Parliament; organize small speaking tours and meetings with the press, politicians, and other activists. KWIA also provides travel funds for tribal representatives to the UN Working Group on Indigenous Peoples in Geneva. As a member of NCOS, the Flemish umbrella organization of Third World support groups, KWIA has valuable contacts with foundations active in the Amazon.

Resources: *TRIBBAL NIEUWS* monthly.

Contact: Wendel Trio
Chairman
KWIA
Lange Lozanastraat 14
2018 Antwerpen
Belgium
Phone: (3) 322-6868
Fax: (3) 322-1870

▶ **Flemish Youth Federation for Environmental Conservation "Natuur 2000"**

Objectives: The Federation is a 1,500-member youth organization, which is a member of the International Youth Union for the Conservation of Nature, an environmental lobbyist and education organization.

Programs: In 1989, the Federation helped collect 600,000 signatures to "Save the Tropical Rainforest," which were handed over to the Minister of the Environment in Brussels and to the Belgium's UN representative. In 1990, the group launched the campaign, calling on local authorities to ban tropical hardwood imports.

Resources: A bimonthly bulletin; the Environmental Information and Education Center for Young People.

Contact: Bart Philips
Secretary General
Flemish Youth Federation
Bervoetstraat 33
B-2000 Antwerpen
Belgium
Phone: (3) 231-2604
Fax: (3) 233-6499

EUROPE
DENMARK

▶ **Resource Groups**

▶ **International Work Group for Indigenous Affairs (IWGIA)**

IWGIA is an independent organization which supports indigenous peoples in their struggle against oppression. For more than ten years, IWGIA has consistently published the widest range of reports on indigenous groups, and their movements.

Resources: The quarterly journal *IWGIA Documentation Series* (in Spanish and English); and the quarterlies *IWGIA Newsletter* (English), and *IWGIA Boletín* (Spanish).

Contact: IWGIA
Fiolstræde 10
DK-1171 Copenhagen K
Denmark
Phone: (33) 124-724
Fax: (33) 147-749

EUROPE
ENGLAND

Angotere-Secoya people of Peru.
Photo: ©Alice Levey

▶ **Amnesty International (AI)**

Objectives: AI has a world-wide membership of one million that works for the release of prisoners of conscience, campaigns for the abolition of torture and the death penalty, and opposes extra-judicial execution by governments.

Programs: In Brazil, AI has been conducting a campaign against rural violence by urging the government to investigate the killings of trade union leaders and their advisers as well as abuses against Amazonians. AI provides information to interested individuals and will take on cases that fall within its mandate.

Resources: The monthly *Amnesty International Newsletter;* the special report, *Brasil Documento.*

Contact: Patricia Feeney
Researcher
Amnesty International
Secretariat
1 Easton St.
London WC1X 8DJ
England
Phone: (1) 833-1771
Fax: (1) 833-5100
Telex: 28502

▶ **Colonialism and Indigenous Minorities Research/Action (CIMRA)**

Objectives: CIMRA formed in the 1970s to monitor European government and corporate activities affecting the Australian Aboriginals. During the early 1980s, CIMRA broadened into an international radical solidarity group concerned with the impacts of mining on indigenous peoples around the world.

Programs: CIMRA's Amazonian work involves researching the mining activities of British companies, such as British Petroleum and Rio Tinto Zinc, and providing the research and detailed company profiles to interested groups and communities.

Resources: *Indigenous Voice,* a two-volume book set on indigenous struggles to protect their lands and cultures.

Contact: Roger Moody or
Angie Aldridge
Coordinators
CIMRA
218 Liverpool Rd.
London, N1 ILE
England
Phone: (1) 609-1852

▶ Friends of the Earth Ltd. (FOE)

Objectives: With 180,000 supporters and 280 local groups, FOE is England's leading environmental organization and the largest of the 38 national FOE groups around the world.

Programs: As part of the Tropical Forest Campaign, FOE has been fundraising for the Kayapó and Macuxí Amazonian Indians to save their lands from destructive development projects and lobbying the Brazilian government for sustainable development alternatives to cattle ranching, logging, and burning. FOE was the first environmental group to produce a detailed analysis of its own country's timber imports and to conduct a national consumer boycott of tropical hardwoods.

FOE helps tribal and environmental groups from the Amazon, particularly from Brazil, by setting up formal meetings with government, bank, and aid agency officials; speaking tours; media outreach; and network building in UK and Europe.

Resources: *Tropical Rainforest Times* newsletter; *Earth Matters* (FOE newsletter); urgent action bulletins, briefings, leaflets.

Contact: Simon Counsell
Rainforest Campaigner
FOE Ltd.
26-28 Underwood Street
London N1 7JQ
England
Phone: (1) 490-0336/1555
Fax: (1) 251-0818 or 490-0881
Greenet: FOETRF

▶ Gaia Foundation

Objectives: Gaia (Greek for Earth) formed in 1984 to "increase our understanding of the Earth as a living whole" by supporting projects and individuals who "regenerate the life of the planet."

Programs: Gaia began working in Brazil by supporting the work of environmentalist José Lutzenberger. In 1988, Gaia formed the Forest People's Support Group (FPSG) to help forest peoples, such as Indians and rubber tappers, protect their homes and cultures.

FPSG has raised funds for the Kayapós' conference in Altamira in February 1988; a communications system for the Union of Indigenous Nations (UNI); legal support for Indian land rights; reclamation schemes for deforested areas; educational programs for Indian university students; and an international exhibition of Kayapó cosmology.

FPSG also helps peoples of the forest build contacts and support in Europe by organizing speaking tours, fundraisers, and other events.

Resources: Videos; the book *A Tribute to the Forests Peoples of Brazil*

Contact: Liz Hosken
Gaia Foundation
18 Well Walk Foundation
Hampstead
London NW3 1LD
England
Phone: (1) 435-5000
Fax: (1) 431-0551

▶ Oxfam UK

Objectives: Oxfam UK is one of six autonomous Oxfam groups around the world that provide financial and technical assistance to grassroots develop projects in the Third World. In the UK, Oxfam builds public support for those projects by speaking in schools and with members of the British government, European Community, and the World Bank.

Programs: Among its many Amazonian projects, Oxfam UK has funded leadership training programs for the Confederation of Indian Nationalities of the Ecuadorian Amazon (CONFENAIE); the establishment of community markets in the Quichua Indian areas of the Ecuadorian Amazon; and organizational development for the Amazonian Central Organization of Indigenous Peoples of Eastern Bolivia (CIDOB). Oxfam UK is also supporting the training of health workers in Brazil in communties dependent on traditional and inexpensive medical care. The program focuses on the use of herbal medicines and techniques such as acupressure and acupuncture.

Resources: *Oxfam News* newsletter; the report *Oxfam's Work in the Amazon Basin-1989;* project summaries; photos and video library; educational materials for schools and teachers.

Contact: Latin America Desk
Oxfam UK
274 Banbury Rd.
Oxford OX2 7DZ
England
Phone: (865) 56777
Telex: 83610
Greenet: oxfam hq

FOE staff and affiliates from around the world at the annual meeting in 1989.
Photo: Juliette Majot.

Survival International Secretariat

Objectives: Founded in 1969, Survival is the leading international organization devoted to supporting tribal peoples' rights to survival and self determination; ensuring that their interests are properly represented in all decisions affecting their futures; securing their ownership and use of adequate natural resources; and seeking recognition of their rights to traditional lands. Survival has supporters in more than 60 countries. The International Secretariat in the U.K. is responsible for organizational research, publishing, campaign planning. Through its Urgent Action Bulletins, Survival is able to reach thousands of people around the world to support critical tribal issues.

Programs: Survival currently handles about 50 cases worldwide, including the Brazilian Amazon. It has launched a major campaign calling on agencies funding Carajás mining operations to institute an emergency programs to safeguard the lands and livelihoods of 13 affected tribes. Survival appealed to the authorities to suspend construction of the Acre Road extension until impacts on Indians were addressed. Survival also joined Indians in efforts to stop oil exploration in the Javari Valley; to lobby for Indian land rights at the constitutional convention; and to cancel plans to build two hydroelectric dams on the Xingu river.

Most recently, Survival has launched a campaign to protect the Yanomami, a tribe whose lands and health have been critically eroded by the invasion of gold miners. At the request of the tribe, Survival's campaign will focus on securing all the Yanomami territory, stopping gold mining, and raising funds for emergency medical work.

Resources: The newsletter *Survival International News* (in English, French, Spanish, Italian) three times/yr.; *Urgent Action Bulletins*, 10 to 12/yr.; a series of special documents and reports including *Bound in Misery and Iron: The Impact of the Grande Carajás Programme on the Indians of Brazil* and *The Amuesha-Yanachaga Project in Peru*; information packets, books, films, videos and slide shows.

Contact: Survival International
310 Edgware Road
London W2 1DY
England
Phone: (1) 723-5535
Fax: (1) 723-4059
Greenet: SURVIVAL

Yanomami Family. Photo: ©Victor Englebert/Survival International

Resource Groups

Norman Myers

Dr. Norman Myers is an independent scientist and consultant on deforestation rates, all outputs of tropical forests ("both goods and services, not just timber and beef"), environmental functions, wildlife and genetic resources, and climatic connections. Dr. Myers is now preparing an update of his 1980 National Academy of Sciences survey of deforestation rates that will emphasize Amazonia.

Resources: The book *The Primary Source*.

Contact: Dr. Norman Myers
Upper Meadow, Old Rd.
Headington
Oxford OX3 8SZ
England
Phone: (865) 750-387
Fax: (865) 741-538

Oxford Forestry Institute

The Institute provides education and training, research, a library, and a information and advisory service on all aspects of forestry, particularly tropical. In the Amazon, the institute manages the British government's collaborative research programs on forestry and the environment.

Resources: Annual report, *Proceedings of Forum on the Future of the Tropical Rainforest*.

Contact: Dr. J. Burley
Director
Oxford University
South Parks Road
Oxford OX1 3RB
England
Phone: (865) 275-000
Fax: (865) 270-708

Royal Botanic Gardens, Kew (RBG)

The RBG was founded in 1840 to conduct research, educational, and horticultural projects. Much of RBG's taxonomic and exploration work is done in rainforest countries in collaboration with government agencies and universities.

RBG's Brazilian activities include coordinating a technical assistance program for Britain's Overseas Development Administration on sustainable uses of the Amazon forest such as establishing a biological reserve in the Caxiuanã National Forest in Pará.

Resources: Botanical science publications: *Kew Bulletin*, *Kew Magazine*, and *Kew Index of Taxonomic Literature*.

Contact: Dr. Ghillean T. Prance
Director
RBG
Kew, Richmond
Surrey TW9 3AB
England
Phone: (1) 940-1171
Fax: (1) 948-1197

EUROPE

HUNGARY
ITALY

Royal Geographical Society (RGS)

RGS is a 10,000-member organization that was founded in 1830 to advance geographical science and improve and diffuse geographical information. In 1987-1988, the RGS organized the Maracá Rainforest Project on the ecological station of Maracá Island, Roraima, Brazil in collaboration with the National Institute of Amazon Research (INPA) in Manaus. Staffed by 150 Brazilian and British scientists, the project is surveying the ecosystem of the Maracá reserve and studying forest regeneration, soils and hydrology, medical etymology, and land development. RGS has also provided financial support for Funatura, Fundacao Brazileira, and Commitee for the Creation of a Yanomami Park (CCPY).

Resources: The Maracá survey report and book.

Contact: Fiona Watson
RGS
1 Kensington Gore
London SW7 2AR
England
Phone: (1) 589-5466
Fax: (1) 581-9918

The Wadebridge Ecological Centre

Started by environmentalist Teddy Goldsmith, Wadebridge publishes *The Ecologist*, a trailblazing magazine whose in-depth investigations into multilateral development bank projects, especially mega hydroelectric dams, helped call international attention to destructive development strategies in the Third World. In 1989, Goldsmith and several other environmental groups gathered 3 million signatures on a petition calling for a special UN session on tropical rainforests and delivered them in wheelbarrels to UN headquarters in New York.

Resources: *The Ecologist* (6 times/yr), special issues include: "Save the Forests, Save the Planet—A Plan for Action" (1987), and "Development and the Amazon" (1989); the two-volume book *The Social and Environmental Effects of Large Dams*.

Contact: Nicholas Hildyard
Wadebridge Ecological Centre
Worthyvale Manor
Camelford
Cornwall PL32 9TT
England
Phone: (258) 73-476
Fax: (258) 72-372

World Youth Rainforest Action

Objectives: The World Youth Rainforest Action was formed in October 1989 by eight European youth groups to fight for the preservation of the world's rainforests.

Programs: The Action's first project will be the international youth campaign to "Save the Amazon." The campaign—the organizational platform calling for sustainable development of the Amazon—was established at the October 1989 meeting. Future activities will be planned in 1990.

Contact: Mauro Porto
c/o World Federation of Democratic Youth
P.O. Box 147
Budapest 1389
Hungary
Phone: 1154-094
Fax: (1) 1352746
Telex: 227197 DIVSZ H

Amici Della Terra Italia (FOE)

(Friends of the Earth/Italy)

Objectives: FOE is Italy's most effective environmental lobbyist group.

Programs: For the last three years, FOE has been lobbying the Italian government, the European Community (EC), and the World Bank to stop funding some of Brazil's most destructive development projects: Xingu hydro dams, Polonoroeste resettlement, and Carajás mining operations. FOE organized Kayapó Chief Paiakan's first speaking tour of Italy (1988) to campaign against the Xingu dams, and helped organize the Kayapó's international conference in Altamira, Brazil.

In May 1989, FOE held the "International Meeting with the Peoples of the Amazon Forest" (Milan), attended by Amazonian Indian representatives, rubber tappers, Brazilian scientists, politicians, environmentalists, and Italian and EC officials. It staged Italy's biggest street rally against the destruction of the Amazon, attended by 12,000 people.

In 1990 FOE is circulating petitions directed at the Italian Prime Minister, the presidents of the EC, and the World Bank to stop funding destructive projects in the Amazon, particularly the new Polonoroeste loan. More than 70,000 signatures have already been collected. FOE will be focusing on strengthening communications with the peoples of the forest.

Resources: The booklets, *Amazzonia a Ferro e Puoco* and *Paradiso Perduto*; a video *Da Altamira a Milano* (*From Altamira to Milan*).

Contact: Roberto Smeraldi
Amici Della Terra
Via Del Sudario
35, 00186 Roma
Italy
Phone: (6) 687-5308 or 686-8289
Fax: (6) 654-8610

Campagna Nord/Sud: Biosfera-Sopravvivenza dei Popoli-Debito

(The North-South Campaign: Biosphere, Human Survival, Debt)

Objectives: Campagna formed in 1988 to change the Northern economic and industrial model, which has been causing extraordinary damage to the environment and people of the South.

Programs: Campagna is investigating the activities of Italian businesses and government in the Amazon, which have negative environmental, social, and cultural impacts. The information will be available to Amazonian and European organizations for campaign work.

Contact: Jutta Steigerwald
Researcher
Campagna Nord-Sud
Via S.Maria Dell'Anima, 30
00186 Roma
Italy
Phone: (6) 686-5842
Telex: (51) 933-524

Lelio Basso International Foundation for the Rights and Liberation of Peoples

Objectives: The Foundation is a human rights organization which focuses on specific cases through its Permanent People's Tribunal.

Programs: The Tribunal has already held 14 sessions around the world in the last ten years and plans to hold one in October 1990 in Europe on the conquering of the Brazilian Amazon.

Contact: Tatiana de Miranda Jordão
Administrative Coordinator
Via Della Dogana Vecchia 5
00186 Rome
Italy
Phone: (6) 654-1468
Fax: (6) 654-0945
Telex: 620 248

EUROPE
NETHERLANDS
NORWAY

International Water Tribunal

Objectives: The Tribunal provides every individual or organization with the opportunity to present cases of water misuse to an independent international jury. After hearing evidence from both plaintiffs and defendants, the jury decides how the cases are in conflict with generally accepted ethical principles, international declarations and conventions, and international and national legislation. The verdict is made public.

Kuruaia Indians of Brazil.
Photo: ©Alex DeMoura King

Programs: The First Tribunal, which took place in 1983 on water pollution in northwest Europe, was highly publicized and resulted in several companies reducing their waste disposal into rivers. The Second Tribunal will take place in October 1991 and focus on all types of water misuse in Africa, Asia, and South America including the impacts of Amazonian dams and mines.

Resources: A newsletter will be published from Spring 1990 to Fall 1991.

Contact: Arthur Van Norden
Coordinator
Damrak 831
Amsterdam 1012 LN
The Netherlands
Phone: (20) 240 610
Fax: (20) 228 384

Working Group Indigenous Peoples (WIP)

Objectives: WIP acts as a "mouthpiece" for the Amazonian indigenous federations of Bolivia (CIDOB), Peru (AIDESEP), Colombia (ONIC), Brazil (UNI), and Ecuador (CONFENAIE) in international human rights forums.

Resources: monthly newsletter *Triball Nieuws*.

Contact: Jacques de Korte
Action Coordinator
WIP
P.O. Box 4098
Amsterdam 1009 AB
Netherlands
Phone: (20)938-625

Resource Groups

Both Ends Environment and Development Service for NGOs

Both Ends provides expert assistance to non-governmental organizations (NGOs) in developing countries who need information and funding from Dutch and other European foundations and organizations. Both Ends has helped 300 groups, including 30 in Brazil.

Resources: Guidelines for a project proposal, audio visual materials.

Contact: Theo van Koolwijk
Coordinator
Damrak 28-30,
Amsterdam 1012 LJ
The Netherlands
Phone: (20)-230823 or 261732
Telex: 70890 snm nl

Foreningen for Internasjonales Vannkraftstudier (FIVAS)

(Association for International Hydro-Power Studies)

Objectives: FIVAS tries to prevent environmental damage and social tragedies from hydropower development, mainly by monitoring and fighting against Norwegian and other European water projects around the world. The group has thoroughly documented serious threats to indigenous peoples and tropical rainforests in its 1988 report.

Programs: In 1989, FIVAS focused its Amazonian work on lobbying the World Bank and Norwegian government to stop funding Brazil's Power Sector II loan. FIVAS representatives have also traveled around the Amazon to investigate the water problems.

Resources: Report on hydro-development in the Third World and Norwegian involvement in it.

Contact: Suein Thore Jensen
Information Secretary
FIVAS
P.O. Box 1116, Blindern
N-0317 Oslo 3
Norway
Phone: (47) 245-5753
Fax: (47) 260-1466

EUROPE
SPAIN
SWEDEN

Above and bottom right: Migrants to the Amazon.
Photos: ©Aguirre/Switkes

▶ **Asociación Pro Derechos Humanos De España (APDHE)**
(Association for Human Rights of Spain)

Objectives: APDHE works with other international native and human rights organizations to protect indigenous peoples and their environment.

Programs: In 1988, APDHE spearheaded efforts with other human rights groups to establish the Pro Amazonia-Spain Commission, which would increase public education and support for Amazonian victims of human rights violations.

Resources: The magazine *Derechos Humanos* (Human Rights).

Contact: Luis Miguel Alonso
Secretario Genera
APDHE
José Ortega y Gasset N°77,
piso 2°
Madrid, 28006
Spain
Phone: (91) 402-2312
Fax: 402-8499
Telex: 41045 APDH E

▶ **Friends of the Earth/Sweden (FOE)**

Objectives: FOE Sweden's international work includes lobbying the Swedish government to stop funding destructive development projects in the Third World either directly or through the multilateral development banks.

Programs: FOE Sweden has sent a representative to the Amazon to travel extensively and investigate development projects. FOE's focus has been on analyzing the impacts of Polonoroeste resettlement project in Brazil.

Resources: Report on Polonoroeste.

Contact: Ulf Rasmusson
FOE/Sweden
Nytorgsgatan 21 B
S-116 22 Stockholm
Sweden
Phone: (8) 34-4122 or 40-3796

EUROPE
SWITZERLAND

▶ **Nouvelle Planete**
(New Planet)

Objectives: Nouvelle Planete is a small non-profit organization committed to grassroots organizing, appropriate technology, and linking up Swiss and Third World groups on issues of rainforest preservation, native rights, and sustainable developement.

Programs: Planete twins villages with towns in Switzerland to those in developing countries to collaborate on some small-scale community development project. This year Planete is working with the Coordinating Body of the Indigenous Organizations of the Amazon Basin (COICA) to start twinning projects with South American Indian communities focusing on land titling.

Contact: Jeremy Narby
Nouvelle Planete
CH-1042 Assens
Switzerland
Telex: MA 40989

EUROPE
WEST GERMANY

▶ **Kampagne für das Leben in Amazonien**
(Campaign for Life in the Amazon)

Objectives: Kampagne is an educational and lobbyist organization on issues related to rainforest destruction and the oppression of forest peoples.

Programs: Kampane has helped Indian delegations from the Amazon obtain access to German politicians, scientists, media, and the public. It is now pressuring the European Parliament on the human rights and environmental problems associated with the gold mining on Yanomami Indian lands in Brazil.

Resources: Newsletter.

Contact: Claudio/Christine Moser
Kampagne für das
Leben in Amazonien
Lotharstr. 14
Bonn 5300
West Germany
Phone: (228) 214-366

▶ **Gesellschaft für Bedrohte Völker (GBV)**
(Society for Endangered Peoples)

Objectives: GBV works in defense of endangered peoples' rights to security, development, religious and cultural identity.

Programs: The group helps gather public support for Amazonian issues by issuing press releases, action alerts, etc.

Resources: The journal *Pogrom*.

Contact: Robert Lessmann
Secretary for Latin America
Gesellschaft für Bedrohte Völker
P.O. Box 2024
3400 Gottingen
West Germany
Phone: (551) 55-822 or 55-823
Telex: 175518101 GfbVGoe.

▶ **Rettet Den Regenwald e.V.**
(Save the Rainforest)

Objectives: Established in 1986, Rettet is a 1,400-member group that maintains an information and action center on rainforest issues.

Programs: Rettet's main activities are publishing journals, and launching letter-writing campaigns, tropical timber boycotts, and demonstrations against stores selling tropical timber and the Carajás mining project in Brazil.

Resources: The quartely *Regenwald Report*.

Contact: Reinhard Behrend
Director
Rettet Den Regenwald
Pöseldorfer Weg 17
2000 Hamburg 13
West Germany
Phone: (40) 410 38 04

▶ **Arbeitsgemeinschaft Regenwald und Artenschutz e.V. (ARA)**
(Working Group on Rainforest and Species Conservation)

Objectives: ARA is an environmental education and lobbyist organization that is one of West Germany's leading rainforest preservation groups.

Programs: ARA investigates projects in the Amazon that are funded or carried out by German companies, banks, and aid agencies and builds public support to stop them.

Resources: *Rainforest Memorandum* on West Germany's role in rainforest destruction; brochures on the rainforest ecology and the timber trade.

Contact: Wolfgang Kuhlman
Director
ARA
P.O. Box 531
D-4800 Bielefeld-1
West Germany
Phone: (521) 60-072
Fax: (524) 62-643

Charcoal production at the Greater Carajás Project.
Photo: ©Aguirre/Switkes

ASIA/PACIFIC
AUSTRALIA INDIA

▶ **Rainforest Foundation of Australia**

Objectives: The Foundation is the Australian chapter of the Rainforest Foundation, an international organization committed to fundraising for Brazilian Indians' campaigns to protect their lands from destruction.

Programs: See Rainforest Foundation (U.S.A.) page 73.

Contact: Dr. Jonathan King
Executive Director
Rainforest Foundation of Australia
P.O. Box 123
Palm Beach
Sydney 2108
Australia
Phone: (2) 973-1438
Fax: (2) 918-8388

▶ **Rainforest Information Centre (RIC)**

Objectives: RIC was started in 1979 by environmentalist John Seed to stop destruction of the world's last remaining rainforests through non-violent demonstrations and public education. RIC is also an ashram where activists can live whether they choose to work as a collective or autonomously.

Programs: RIC provides information to and launches letter-writing and petition campaigns with the more than 50 rainforest action groups throughout Australia on various issues related to saving the Amazon.

Resources: The quarterly *World Rainforest Report*.

Contact: Mara T.
Coordinator
RIC
P.O. Box 368
Lismore 2480, NSW
Australia
Phone: (66) 218-505

▶ **Resource Groups**

▶ **Research Foundation for Science and Ecology**

The Foundation studies and monitors the role of international aid and corporations in the destruction of the world's rainforests, including the Amazon. It also scrutinizes from a Third World perspective the large-scale plans by the World Bank and international wildlife organizations to "save the rainforest," such as the Tropical Timber Action Plan and the new Biodiversity Action Plan.

Programs: Critiques of the Tropical Timber Action Plan and Biodiversity Action Plan.

Contact: Dr. Vandana Shiva
Research Foundation for Science and Ecology
105 Rajpur Rd.
Dehea Inn
UP, India
Phone: (135) 23-374

ASIA/PACIFIC
JAPAN/MALAYSIA

▶ **Nettairin Kodo Network (JATAN)**

(Japan Tropical Forest Action Network)

Objectives: JATAN was formed in 1987 by Japan's leading environmental, human rights, women's, and consumer groups to conduct research, campaign, and lobby in Japan for rainforest preservation. JATAN's primary goal is to reform Japanese timber-trade and development polices, which promote destruction of rainforest.

Programs: JATAN is now identifying Japanese corporations and funding agencies (JICA, OECF, EXIM Bank.) which are involved in development projects in the Amazon such as Xingu dams, Acre road construction, and Carajás mines. JATAN will use the information to lobby the appropriate members of the Japanese Diet and provide relevant data to Brazilian organizations. JATAN helps organize speaking tours, and meetings with Japanese officials and press for visiting Amazonian activists.

Resources: *JATAN News* (in Japanese) and the study *Timber from the South Seas: An Analysis of Japan's Tropical Timber Trade and its Environmental Impact*.

Contact: Yoichi Kuroda
Coordinator
JATAN
50/Shinwa Bldg.
9-17 Sakuragaoka
Shibuya-ku Tokyo 150
Japan
Phone: (3) 770-6308
Fax: (3) 770-0727

▶ **Resource Groups**

▶ **Third World Network Features**

This news network translates and distributes articles on rainforests and other international environmental issues to nonprofit and alternative publications world-wide. The network is part of the World Rainforest Movement, which is made up of rainforest organizations around the world.

Contact: Martin Khor Kok Peng
Third World Network
87 Cantonment Rd.
10250 Penang
Malaysia
Phone: (4) 373-511 or 373713
Fax: (4) 368-106
Telex: MA 40989

Recommended Books and Films

ADVENTURE

1. *The Rio Tigre and Beyond,* by F. Bruce Lamb (North Atlantic Books, 1985, $12.95). Experiences with the hallucinogenic forest vine *ayahuasca*; appendix of medicinal plants (see *Wizard of the Upper Amazon*).
2. *Running the Amazon,* by Joe Kane (Alfred Knopf, 1989, $19.95). National bestselling account of a kayak expedition from source to mouth of the Amazon River.
3. *Wizard of the Upper Amazon,* by F. Bruce Lamb (North Atlantic Books, 1975, $6.75). Kidnapped by Indians while working on a remote Amazonian river, Manuel Cordova is trained as a shaman.

DEVELOPMENT

1. *The Debt Squads: The U.S., the Banks, and Latin America,* by Sue Branford and Bernardo Kucinski (Zed Books, 1988, $49.95/$12.95). Argues that the Third World debt cannot and should not be paid.
2. *Economics as if the Earth Really Mattered,* by Susan Meeker-Lowry (New Society Publishers, 1988, $34.95/$9.95). Ethical investing for social change.
3. *The Fate of the Forest: Developers, Destroyers and Defenders of the Amazon,* by Susanna Hecht & Alexander Cockburn (Verso, 1989, $24.95). Provocative book recounting history of Brazilian Amazon development, with a special emphasis on the activism of grassroots movements.
4. *A Fate Worse Than Debt,* by Susan George (Grove Press, 1988, $17.95/$8.95). International debt and its relationship to social and environmental problems in the Third World.
5. *Fight for the Forest: Chico Mendes in His Own Words,* by Chico Mendes (Latin America Bureau, 1989, $6.95). A collection of interviews with the rubber tapper leader, describing the evolution of the *seringueiros'* movement.
6. *Hoofprints on the Forest,* by Douglas Shane (Ishi, 1986, $27.50). Impacts of cattle ranching in Latin American rainforests.
7. *Indigenous People and Tropical Forests,* by Jason Clay (Cultural Survival, 1988, $8.00). Case studies of sustainable development strategies of indigenous people.
8. *Indigenous Peoples: A Field Guide for Development,* by John Beauclerk and Jeremy Narby with Janet Townsend (Oxfam, 1988, £19.95/£4.95). General insights into practical strategies for working with indigenous peoples, from the experiences of Oxfam UK.
9. *Saving the Tropical Forests,* by Judith Gradwohl and Russell Greenberg (Island, 1988, $24.95). Strategies for protecting rainforests in various countries, including the Amazon.
10. *The Last Frontier: Fighting for Land in the Amazon*, by Susan Branford and Oriel Glock (Zed Books,1985, $39.95/$12.25). Account of land conflicts on the frontier, especially involving small farmers and squatters.
11. *The Social and Environmental Impacts of Large-Scale Hydroelectric Dams,* by Edward Goldsmith and Nicholas Hildyard (Sierra Club, 1984, $29.95). Detailed indictment of devastation caused by high dams.

FICTION

1. *Emperor of the Amazon,* by Márcio Souza (Avon, 1980, $2.95). Hilarious tale of how Brazil expropriated Acre state from Bolivia during the rubber boom.
2. *Mad Maria,* by Márcio Souza (Avon, 1985, $4.95). Madcap, satirical story of the construction of the Madeira-Mamoré, the "devil's railway," in turn-of-the century Rondônia.
3. *Maíra,* by Darcy Ribeiro (Random House,1983, $7.95). Novel by a leading Brazilian anthropologist tells of relationship between a nurse and a non-traditional Indian in the Amazon.
4. *At Play in the Fields of the Lord, by* Peter Matthiessen (Random House, 1987, $6.95). Novel about evangelical missionaries' activities with native people in the Amazon.
5. *The Storyteller,* by Mario Vargas Llosa (Farrar, Strauss & Giroux, 1989, $17.95). Vivid story of the kinship between a Jewish outcast and the Machiguenga Indians of the Peruvian Amazon.

GENERAL, DESCRIPTIVE

1. *In the Rainforest*, by Catherine Caulfield (University of Chicago Press, 1986, $11.95). Best journalistic work on the rainforest issue, with descriptions of environments and threats to them, including the Amazon.
2. *Jacques Cousteau's Amazon Journey,* by Jacques-Yves Cousteau and Mose Richards (Harry N. Abrams, 1984, $39.95). Large format photo book about an exciting expedition from source to mouth of the Amazon.
3. *Lessons of the Rainforest,* edited by Suzanne Head and Robert Heinzman (Sierra Club, San Francisco, $24.95/$14.95). Causes, consequences, and solutions to tropical deforestation as seen by 24 well-known conservationists.
4. *This Place is Wet,* by Vicki Cobb; illustrated by Barbara Lavallee (Walker, 1989, $12.95). Intelligent, interesting childrens' book, serving as an introduction to the Amazon for ages 7 to 9.
5. *The Primary Source,* by Norman Myers (W.W. Norton, 1984, $10.95). Biodiversity and the danger of species extinction.
6. *Tropical Rainforests: Endangered Environment,* by James Nations (Franklin Watts, 1988, $12.90). Comprehensive description of value of rainforests, particularly accessible to junior high and high school students.

NATURAL ENVIRONMENT

1. *Amazonia: Man and Culture in a Counterfeit Paradise,* by Betty Meggers (Harlan Davidson, 1971, $10.95). Seminal work on the fragility of the Amazon ecosystem and aboriginal adaptations to its ecological limits.
2. *The Deluge and the Ark: A Journey into Primate Worlds,* by Dale Peterson (Houghton Mifflin, 1990, $24.95). Endangered primates in the rainforests of South America, Africa, and Asia.
3. *Man, Fishes, and the Amazon,* by Nigel Smith (Columbia University, 1981, $22.50). How fish and river life of the Amazon are managed by riverine people.

4. *Margaret Mee: In Search of Plants of the Amazon Rainforest,* edited by Tony Morrison (Nonesuch Expeditions, 1988, $39.95). Stunningly beautiful botanical drawings and photos by an English naturalist. Expensive but well worth the investment.
5. *A Narrative of Travels on the Amazon & Rio Negro,* by Alfred R. Wallace (Greenwood, 1969 reprinted from 1889, $35.00). A 19th century naturalist in the Amazon.
6. *The Naturalist on the River Amazon,* by Henry Walter Bates (Penguin, 1989, $8.95). Tales of a pioneering 19th century botanist.
7. *Plants of the Gods,* by Richard Schultes and Albert Hofmann (Mc-Graw-Hill, 1979, $34.95/$16.95). Illustrated survey of plants considered sacred in various cultures.
8. *The Rivers Amazon,* by Alex Shoumatoff (Sierra Club Books, 1986, $8.95). Botanist and journalist describes trip to the rainforest.
9. *Tropical Nature: Life & Death in the Rain Forests of Central and South America,* by Adrian Forsyth & Ken Miyata (Charles & Scribner's Sons, 1987, $16.95/$7.95). Descriptions of wonders of nature in the Americas' rainforests.

HISTORY

1. *Amazon Frontier,* by John Hemming (Harvard University Press, 1987, $29.95). Sequel to *Red Gold*; describes the continuing interaction between Brazilian natives and the advancing frontier.
2. *Prehistoric America: An Ecological Perspective,* by Betty Meggers (Aldine de Gruyter, 1979, $29.95/$15.95). Comprehensive archaeological survey of the Americas, including Amazonia.
3. *Red Gold,* by John Hemming (Harvard University Press, 1978, $38.00). Account of the first 200 years of the conquest of Brazilian Indians; well-documented and chilling.

INDIGENOUS PEOPLES

1. *Amazonian Cosmos,* by Geraldo Reichel-Dolmatoff (University of Chicago, 1974, $17.95). Study of mythology and shamanism in Rio Negro region of Colombia.
2. *Before the Bulldozer: The Nambiquara Indians & The World Bank,* by David Price (Seven Locks Press, 1989, $17.95). An anthropologist attacks World Bank's lack of commitment to protection of native peoples impacted by the Polonoroeste Project.
3. *Fishers of Men or Founders of Empire? The Wycliffe Bible Translators in Latin America,* by David Stoll (Zed, 1983, $29.50/$12.50). Causes and effects of evangelical missionaries' interactions with native cultures.
4. *Peoples of the Tropical Rainforest,* edited by Christine Padoch and Julie Sloan Denslow (University of California, 1988, $37.50). Large-format photo book with collection of articles by scholars on Indians and riverine people of the Amazon.
5. *The Savage and the Innocent,* by David Maybury-Lewis (Beacon, 1965, $11.95). Founder of Cultural Survival gives intriguing insights from his fieldwork among the Xavantes of the Brazilian Amazon.
6. *Victims of the Miracle,* by Shelton Davis (Cambridge University Press, 1977, $12.50). Excellent account of how Amazonian Indians have been affected by activities of multinational corporations.

FILMS

1. *Amazonia: Voices from the Rainforest.* Rosainés Aguirre, Glenn Switkes, producers. 80 min. Strategies of the people of the Amazon to defend their homeland, scheduled for release in Fall 1990. Contact: AMAZONIA, P.O. Box 10044, Oakland, CA 94610. Phone: (415) 243-4146.
2. *Amazonia: A Celebration of Life.* Andrew Young, producer. 23 min., 1983. Diversity of wildlife in the Amazon, produced in association with World Wildlife Fund. State University at Stony Brook, NY. Phone: (516) 632-6484.
3. *Banking on Disaster.* Adrian Cowell, producer. 78 min., 1989. Part of the "Decade of Destruction" series produced by Central Television of London on the environmental devastation in Rondônia and Acre, Brazil. Bullfrog Films, Oley, PA 19547. Phone: (1-800) 543-FROG.
4. *The Brazilian Connection.* Helena Solberg-Ladd and the International Womens' Film Project, producers. Video, 60 min., 1983. Impacts of Brazil's foreign debt crisis. Still timely. Cinema Guild, 1697 Broadway, Suite 802, New York, NY 10019. Phone: (212) 246-5522.
5. *Chico Mendes: A Voice from the Amazon.* Miranda Smith, director. 60 min., 1989. Documentary on the life and work of Chico Mendes, including interviews with the rubber tapper leader conducted shortly before his murder. Cinema Guild (see above).
6. *Contact: The Yanomami Indians of Brazil.* Geoffrey O'Connor, producer. 28 min., 1985. A look at the impacts of the invasion of Yanomami territory by 45,000 gold miners. Realis Pictures, 32 Union Square East, Suite 816, New York, N.Y. 10003. Phone: (212) 505-1911.
7. *Ikatena: Vamos Caçar.* Luis Paulino dos Santos, producer. 20 min., 1985. Beautiful Brazilian documentary on children of the Zoró tribe of Rondônia, who learn to hunt by imitating their fathers and animals. Transa International Films, 780 Second Avenue, San Francisco, CA 94118. Phone: (415) 221-1783.
8. *Iracema.* Jorge Bodansky, director. 90 min., 1975. Fiction film that, perhaps better than any documentary, captures the feeling of the Transamazon Highway, told through the adventures of an Indian girl who supports herself as a prostitute. Cinema Guild (see above).
9. *Land of the Indians.* Zelito Viana, director. 120 min., 1979. Brazilian film about the situation of the Indians. Still not out-of-date, it is a thoughtful study of indigenous rights. New Yorker Films, 16 West 61st St., New York, NY 10023. Phone: (212) 247-6110.
10. *Listen Caraca.* Carlos Azpurua, producer. 19 min., 1979. Testimony of a Yecuana chief on the impacts of evangelical missionaries on his tribe. Cinema Guild (see above).
11. *Runa: Guardians of the Forest.* Ellen Speiser and Dominique Irvine, producers. 27 min., 1990. The Quichua Indians of Ecuador use their traditions to manage the resources of the rainforest. Write: 257 Fair Oaks St., San Francisco, CA 94110. Phone: (415) 826-8432.
12. *The Sound of Rushing Water.* Ricardo Tankamash and Bruce Horowitz, producers. 42 min., 1973. An important film, because it was made in conjunction with the Shuar Indians of Ecuador, who talk about their culture and the formation of the Shuar Federation. Cinema Guild (see above).

Organizational Index

ABRA, 48
Abya-Yala Editions and Cultural Center, 57
Ação Democrática Feminina Gaúcha— Amigos da Terra, 39
Ação Ecológica Vale do Guaporé, 39
Acción Ecológica, 54
ACIPY, 62
ADFG, 39
Ad Hoc Committee of Nut Gatherers, 28
AGAPAN, 39
AGEN, 48
Agência Ecumênica de Notícias, 48
AIDESEP, 58
Aid for the Indigenous People of Eastern Bolivia, 29
Aliança dos Povos da Floresta, 32
Alliance of the Peoples of the Rainforest, 32
Amazon Center of Anthropology and its Practical Applications, 60
Amazon Project, 64
Amazon Research and Training Program, 75
AME, 39
Amerindian Research Center, 58
Amici Della Terra Italia, 81
Amnesty International, 78
Amnesty International USA, 66
Andean Judges' Commission, 52
APCOB, 29
Aragua Conservation Society, 64
Araracuara Corporation, 53
Arbeitsgemeinschaft Regenwald und Artenschutz e.V., 84
ARC, 58
ARCIBRN, 32
ASEO, 29
Ashoka: Innovators for the Public, 75
Asociación Ametra 2001, 59
Asociación Civil Indígena de los Pueblos Yucpa, 62
Asociación de Ecología y Conservación, 59
Asociación Ecológica del Oriente, 29
Asociación Interétnica del Desarrollo de la Selva Peruana, 58
Asociación Pro Derechos Humanos De España, 83
AS-PTA, 39
Assessoria e Serviços a Projetos em Agricultura Alternativa, 39
Associação Brasileira de Reforma Agrária, 48
Associação Das Comunidades Remanescentes de Quilombos do Município de Oriximiná, 35
Associação Gaúcha para a Proteção do Ambiente Natural, 39
Associação de Informações da Amazônia, 39
Associação de Mulheres do Alto Rio Negro (Numiã-Kuruã), 32
Associação Mundial de Ecologia, 39
Associação Rural das Comunidades Indígenas do Baixo Rio Negro, 32
Association of Communities Remaining from Colonies of Escaped Slaves in the Municipality of Oriximiná, 35
Association for Endangered Peoples-Austria, 77
Association for Human Rights of Spain, 83
Association for International Hydro-Power Studies, 82
Association of Women of the Upper Rio Negro, 32
Atingidos de Cachoeira Porteira, 36
Atingidos de Ji-Paraná, 36
Audubon Conservation Society of Venezuela, 64
Ayuda para el Campesino Indígena del Oriente Boliviano, 29

Bank Information Center, 66
Better World Society, 75
BIOMA, 63
Bolívar Ecological Group, 64

Both Ends Environment and Development Service for NGOs, 82
Brazilian Association for Agrarian Reform, 48
Brazilian Foundation for the Conservation of Nature, 44
Brazilian Institute of Social and Economic Analyses, 49
Brazil Network, 66
British Tropical Agricultural Mission, 31

CAAAP, 60
CAISET, 64
Campagna Nord/Sud: Biosfera-Sopravvienza dei Popoli-Debito, 81
Campaign for Life in the Amazon, 84
Campanha Nacional pela Defesa e Desenvolvimento da Amazônia, 39
Campanha Nacional pela Reforma Agrária, 40
Caribbean Organization of Indigenous Peoples, 57
Casa do Estudante Autóctone do Rio Negro, 33
Catalyst, 66
CATU, 36
CCIM, 28
CCPY, 42
CDC, 31, 60
CDDH, 40
CDDH, 56
CEACON, 41
CEARN, 33
CECAAM, 56
CEDI, 40
CEDIA, 60
CEDIME, 56
CEDISH, 57
CEDLA, 31
CEDOIN, 31
CEES, 53
Center for Advanced Amazon Studies, 50
Center of Amazonian Research and Promotion, 60
Center of Amazon Theological Studies, 60
Center of Audiovisual Communications in Amazonia, 56
Center in Defense of Human Rights, 40
Center for the Development of Indigenous Amazonia, 60
Center of Documentation and Information, 31
Center of Documentation and Information on the Social Movements of Ecuador, 56-57
Center of Education, Research, and Advice for Unions and Popular Movements, 48
Center of Environmental Filming, 41
Center for Indigenous Education, Promotions and Self-sufficiency, 62
Center for Indigenous Knowledge for Agriculture and Rural Development, 75
Center of Moxeño Indian Councils, 28
Center of Research and Promotion of Peasants, 30
Center for Research and Regional Development, 52
Center of Service for Popular Action, 63
Center of Studies and Activities for the Conservation of Nature, 41
Center of Studies and Promotion of Development, 60
Center for the Study of Labor and Agrarian Development, 31
Central de Cabildos Indígenas Moxeños, 28
Central Obrera Departamental, 28
Central Organization of Indigenous Peoples of Eastern Bolivia, 28
Central de Pueblos Indígenas del Oriente Boliviano, 28
Central Workers' Organization, 28
Centro de Comunicaciones Audiovisuales de la Amazonia, 56
Centro de Datos para la Conservación, 31, 60
Centro de Defesa dos Direitos Humanos, 40
Centro para el Desarrollo de la Amazonía Indígena, 60

Centro de Documentación e Información, 31
Centro de Documentación e Información de los Movimientos Sociales del Ecuador, 56-57
Centro de Documentación e Investigación Shuar, 57
Centro Ecumênico de Documentação e Informação, 40
Centro de Educação, Pesquisa e Assessoria Sindical e Popular, 48
Centro de Educación, Promoción y Autogestión Indígena, 62-63
Centro de Estudios para el Desarrollo Laboral y Agrario, 31
Centro de Estudios de Enfermedades Tropicales Simón Bolívar, 64
Centro de Estudios y Promoción de Desarrollo, 60
Centro de Estudios Teológicos Amazónicos, 60
Centro de Estudos e Atividades de Conservação da Natureza, 41
Centro de Filmagens Ambientais, 41
Centro Interdisciplinario de Estudios Regionales, 52
Centro de Investigación y Documentación para el Desarrollo del Beni, 30
Centro de Investigación y Promoción Amazónica, 60
Centro de Investigación y Promoción del Campesino, 30
Centro Mari de Educação Indígena, 48
Centro de Pastoral Vergueiro, 48
Centro de Trabalhadores da Amazônia, 41
Centro de Trabalho Indigenista, 41
Centro de Servicio de la Accíon Popular, 63
CEPAI, 62-63
CEPASP, 48
CESAP, 63
CETA, 60
Chico Mendes Fund, 75
CIDDEBENI, 30
CIDER, 52
CIDOB, 28
CIMI, 43-44
CIPA, 60
CIPCA, 30
CIR, 33
CMEI, 48
CNDDA, 39-40
CNRA, 40
CNS, 37
COD, 28
COIP, 57
Coletivo de Realizadores de Audiovisuais da Amazônia, 48
Collective of Audiovisual Artists of the Amazon, 48
Colombian Corporation of Social Projects, 53
Colonialism and Indigenous Minorities Research/Action, 78
Comisión Andina de Juristas Seccional Colombiana, 52
Comisión por la Defensa de los Derechos Humanos, 56
Comissão dos Atingidos por Tucuruí, 36
Comissão pela Criação do Parque Yanomami, 42
Comissão em Defesa do Rio Uatumã, 35
Comissão Pastoral da Terra, 42
Comissão Pró-Indio, 43
Comissão Regional de Atingidos Por Barragens, 36
Comissão Regional dos Atingidos pelo Complexo Hidrelétrica do Xingu, 36
Comité Ad Hoc de Zafreros Empatronados de la Castaña, 28
Comitê de Defesa da Ilha de São Luis, 36
Commission for the Creation of a Yanomami Park, 42
Commission for the Defense of Human Rights, 56
Commission in Defense of the Uatumã River, 35
Commission of Victims of Tucuruí, 36
Committee in Defense of the Island of São Luis, 36
CONAGE, 44
CONAIE, 54
CONAP, 58
Confederación de Nacionalidades Amazónicas del Perú, 58
Confederación de Nacionalidades Indígenas de la Amazonia Ecuatoriana, 54
Confederación de Nacionalidades Indígenas del Ecuador, 54
Confederación Sindical Unica de Trabajadores Campesinos de Bolivia, 28
Confederation of Amazon Nationalities of Peru, 58
Confederation of Indian Nationalities of the Ecuadorian Amazon, 54
Confederation of Indigenous Nationalities of Ecuador, 54
Confederation of Union Peasant Workers of Bolivia, 29
CONFENIAE, 54
CONIF, 53
CONIVE, 62
Consejo Indio de Venezuela, 62
Consejo Regional Indígena del Vaupés, 51
Conselho Indigenista Missionário, 43-44
Conselho Indígena de Roraima, 33
Conselho Nacional Dos Seringueiros, 37
Conservation Data Center, 31, 60
Conservation International, 66
Cooperativa Agrícola Integral Campesino, 29
Cooperative of Agricultural Peasants, 29
Coordenação Nacional de Geólogos, 44
Coordinadora de Solidaridad con los Pueblos Indígenas del Oriente Boliviano—Asamblea Permanente de Derechos Humanos, 30
Coordinating Committee in Solidarity with the Indigenous Peoples of Eastern Bolivia—Permanent Assembly on Human Rights, 30
Copal, 61
Corporación Colombiana para la Amazonía Araracuara, 53
Corporación Colombiana de Proyectos Sociales, 53
Corporación de Estudios de Sistemas Ecológicos, Económicos y Sociales, 53
Corporación Grupo Ecológico, 52
Corporación Nacional de Investigación y Fomento Forestal, 53
Corporation of Ecological Systems, Economic and Social Studies, 53
CORPOS, 53
CPI, 43
CPT, 42
CPV, 48
CRAB, 36
CRACOHX, 36
CRAVA, 48
CRIVA, 51
CSUTCB, 29
CTA, 41
CTI, 41
Cultural Survival, 67

Democratic Feminine Action of Rio Grande do Sul, 39
DESCO, 60

Earth Medicine, 76
Ecological Action, 56
Ecological Action for the Guaporé Valley, 39
Ecological Association of Eastern Bolivia, 29
ECOPORÉ, 39
ECOTROPICA, 44
Ecumenical Center For Documentation and Information, 40
Ecumenical News Agency, 48
Ediciones Abya-Yala and Centro Cultural Abya-Yala, 57
Emilio Goeldi Museum of Pará, 50
Environmental Defense Fund, 68
Environmental Defense League, 31
Environmental Policy Institute/Friends of the Earth, 68
Escuela Amazónica de Pintura Usko-Ayar, 61
Esquadrão da Vida Produções Culturais, 49
European Environmental Bureau/Working Group on Amazonia, 77

FAN, 30
FASE, 49
FBCN, 44

FCCNL, 59
FECONAMN, 59
Federação das Organizações Indígenas do Rio Negro, 33
Federação de Orgãos para Assistência Social e Educacional, 49
Federación de Comunidades Campesinas y Nativas de Loreto, 59
Federación de Comunidades Nativas del Medio Napo, 59
Federación de Organizaciones Indígnas de Napo, 55
Federación Nacional Sindical Unitaria Agropecuaria, 51
Federation of Indigenous Organizations of the Rio Negro, 33
Federation of Native Communities of the Middle Napo, 59
Federation of Organizations for Social and Educational Assistance, 49
Federation of Peasant and Native Communities of Loreto, 59
FENSUAGRO-CUT, 51
Flemish Youth Federation for Environmental Conservation, 78
FOIN, 55
FOIRN, 33
For Life, 56
Foreningen for Internasjonales Vannkraftstudier, 82
Foundation of Aid to Life in the Tropics, 44
FPCN, 60
Friends of Nature, 30
Friends of the Earth Ltd., 79
Friends of the Earth/Italy, 81
Friends of the Earth/Sweden, 83
FUNATURA, 45
Fundação, 45
Fundação de Apoio a Vida nos Trópicos, 44
Fundação Brasileira para a Conservação de Natureza, 44
Fundação Gaia, 44
Fundação Pró-Natureza, 45
Fundação de Tecnologia do Estado do Acre, 49
Fundacíon Amigos de la Naturaleza, 30
Fundación Manoa, 52
Fundación Maquipucuna, 56
Fundación Natura, 56
Fundación Peruana para la Conservación de la Naturaleza, 60
Fundación Puerto Rastrojo, 52
Fundación Venezolana para la Conservación de la Diversidad
  Biológica, 63
FUNTAC, 49

Gaia Foundation, 44
Gaia Foundation, 79
Gaúcha Association for the Protection of the Natural Environment, 39
GEA, 52
Gesellschaft für Bedrohte Völker, 84
Gesellschaft für Bedrohte Volker-Osterreich, 77
GHRA, 58
Global Exchange , 68
GREBO, 64
Greenpeace USA, 69
Grupo Ecológico Bolívar, 64
Guyana Human Rights Association, 58

IAMA, 45
IBASE, 49
IDAC, 30
IDEA, 57
IDESP, 49
IDMA, 61
IEA, 46
Indian Council of Venezuela, 62
Indian Law Resource Center, 69
Indigenist Support Center, 41
Indigenous Civil Association of Yucpa Peoples, 62
Indigenous Council of Roraima, 33
INESC, 46
Information Association of Amazônia, 39

INPA, 50
Institute of Amazon Studies, 46
Institute of Anthropology and the Environment, 45
Institute of Assistancefor Popular Law, 45
Institute of Development and the Environment, 61
Institute of Documentation and Support for Peasant/Indian Peoples, 30
Institute of Farming Strategies, 57
Institute for International Cooperation and Development, 75
Institute of Socio-Economic Development of Pará State, 49
Institute of Socio-Economic Studies, 46
Instituto de Antropología e Meio Ambiente, 45
Instituto Apoio Jurídico Popular, 45
Instituto Brasileiro de Análises Sociais e Econômicas, 49
Instituto de Desarrollo y Medio Ambiente, 61
Instituto do Desenvolvimento Econômico-Social do Pará, 49
Instituto de Documentación y Apoyo Campesino, 30
Instituto de Estrategias Agropecuarias, 57
Instituto de Estudos, 46
Instituto de Estudos Amazônicos, 46
Instituto Nacional de Pesquisas na Amazônia, 50
Interethnic Association for the Development of the Peruvian Forest, 58
International Rivers Network, 69
International Society of Tropical Foresters, 75
International Water Tribunal, 82
International Work Group for Indigenous Affairs, 78

Japan Tropical Forest Action Network, 85
JATAN, 85
Joaquim Caetano da Silva Museum, 50

Kampagne für das Leben in Amazonien, 84

Lelio Basso International Foundation for the Rights and Liberation
  of Peoples, 82
LIDEMA, 31
Liga de Defensa del Medio-Ambiente, 31

MAREWA, 46
Mari Center of Indigenous Education, 48
Marirí Ecological Movement, 46
MBPV, 50
Medicina da Terra, 76
Metareila Organization of the Suruí Indian People, 33
Misión Británica de Agricultura Tropical, 31
Missionary Council for Support of Indigenous Peoples, 43-44
Missouri Botanical Gardens, 76
MNDDH, 47
Moiwana, 62
Montreal Native Communications, 65
Movement of Landless Rural Workers, 38
Movement of Support for the Resistance of the Waimiri-Atroari, 46
Movement of Women from the Countryside and the City, 38
Movimento de Apoio para a Resistênca Waimiri-Atroari, 46
Movimento Botucatuense Pró-Vida, 50
Movimento Ecológico Marirí, 46
Movimento de Mulheres Do Campo e da Cidade, 38
Movimento Nacional de Defesa dos Direitos Humanos, 47
Movimento dos Trabalhadores Rurais Sem Terra, 38
MST, 38
Museu Joaquim Caetano da Silva, 50
Museu Paraense Emilio Goeldi, 50
Myers, Dr. Norman, 80

NAEA, 50
National Amazon Research Institute, 50
National Campaign for Agrarian Reform, 40
National Campaign for the Defense and Development
  of Amazônia, 39-40

National Coordinating Commitee of Geologists, 44
National Corporation for Forest Research and Development, 53
National Council of Rubber Tappers, 37
National Farmers Organization, 57
National Indigenous Organization of Colombia, 51
National Movement in Defense of Human Rights, 47
National United Agricultural Labor Federation, 51
National Wildlife Federation, 70
Native Students' House of the Rio Negro, 33
The Nature Conservancy, 70
NDI, 47
Nettairin Kodo Network, 85
New England Tropical Forest Project, 71
New York Botanical Gardens, 76
NFO, 57
North-South Campaign: Biosphere, Human Survival, Debt, 81
Nouvelle Planete, 83
Núcleo de Altos Estudos Amazônicos, 50
Núcleo de Direito Indígena, 47
Núcleo de Estudios de la Amazonía Colombiana, 53
Núcleo de Estudios Huitoto y Muinane, 53
Nucleus of Colombian Amazon Research, 53
Nucleus of Huitoto and Muinane Research, 53
Nucleus of Indigenous Rights, 47

OCIM, 33
ONIC, 51
OPAN, 47
Operação Anchieta, 47
Operation Anchieta, 47
Organização do Conselho Indígena Munduruku, 33
Organização Metareila do Povo Indígena Suruí, 33
Organización Nacional Indígena de Colombia, 51
Organization of the Munduruku Indigenous Council, 33
Oxfam America, 71
Oxfam UK, 79
Oxford Forestry Institute, 80

Pastoral Land Commission, 42
Peruvian Foundation for the Centro Amazónico de Antropología y Aplicación Práctica, 60
Peruvian Society of Environmental Law, 61
Por La Vida, 56
Probe International, 65
Pro-Indian Commission, 43
Pro-Life Movement of Botucatu, 50
Pro-Nature Foundation, 45
Programa Venezolano de Educación-Acción en Derechos Humanos, 64
Project Abraço, 71
Project River Dolphin, 53
Proterra, 61
PROVEA, 64
Proyecto Amazonas, 64

Rainforest Action Network, 72
Rainforest Alliance, 73
Rainforest Foundation, 73
Rainforest Foundation of Australia, 85
Rainforest Futures, 73
Rainforest Health Alliance, 76
Rainforest Information Centre, 85
Regional Commission of Dam Victims, 36
Regional Commission of Victims of the Xingu Hydroelectric Complex, 36
Regional Indigenous Council of the Vaupés, 51
Research and Documentation Center for the Development of the Beni Region, 30
Research Foundation for Science and Ecology, 85

Rettet Den Regenwald e.V., 84
Royal Botanic Gardens, Kew, 80
Royal Geographical Society, 81
Rural Association of Indigenous Communities of the Lower Rio Negro, 32

Save the Rainforest, 84
Scarboro Foreign Mission Society, 65
Shuar Documentation and Research Center, 57
Sierra Club, 73-74
Sierra Club Legal Defense Fund, 73
Simón Bolívar Center for the Study of Tropical Diseases, 64
Sindicato Dos Trabalhadores Rurais, 38
Smithsonian Institution, 76
Sociedad Conservacionista Aragua, 64
Sociedad Conservacionista Audubon de Venezuela, 64
Sociedade pela Preservação dos Recursos Naturais e Culturais da Amazônia, 50
Sociedad Peruana de Derecho Ambiental, 61
Society for Endangered Peoples, 84
Society for the Preservation of Natural and Cultural Resources of Amazônia, 5
Sócio-Econômicos, 46
SORPREN, 50
SOS-AMAZONIA, 45
South and Central American Indian Information Center, 74
SPDA, 61
Squadron of Life Cultural Productions, 49
Stuengroep voor Inheemse Volkeren, 77
Support Group for Indigenous Peoples, 77
Support and Services to Alternative Agriculture Projects, 39
Survival International Secretariat UK, 80
Survival International USA, 74

Taskforce on the Churches and Corporate Responsibility, 65
Technological Foundation of the State of Acre, 49
Third World Network Features, 85
Threshold Foundation, 76
Tunasarapa Suriname, 62

UNI, 34-35
União Das Nações Indígenas, 34-35
União Protetora do Ambiente Natural, 47
Union of Indigenous Nations, 34-35
Union of Rural Workers, Xapuri, Acre, 38
Union to Protect the Natural Environment, 47
Unuma Support Group to Indigenous People, 64
Unuma-Sociedad Civil de Apoyo al Indígena, 64
UPAN, 47
Usko-Ayar Amazonian School of Painting, 61

Venezuelan Foundation for the Conservation of Biological Diversity, 63
Venezuelan Program of Education-Action on Human Rights, 64
Vergueiro Church Center, 48
Victims of Cachoeira Porteira Dam, 36
Victims of Ji-Paraná Dam, 36

Wadebridge Ecological Centre, 81
Wildlife Conservation International, 74
Workers' Center of Amazonia, 41
Working Group Indigenous Peoples, 82
Working Group on Rainforest and Species Conservation, 84
World Association of Ecology, 39
World Resources Institute, 76
World Wildlife Fund & Conservation Foundation, 74
World Youth Rainforest Action, 81
Yanesha Forestry Cooperative, 59
Yanomamo Survival Fund, 75

DELTA COLLEGE LIBRARY

3 1434 00348 8024

DISCARD

OCT 11 2016

GE 160 .A43 A48 1990

Amazonia, voices from the
rainforest   06/2014